Applie

Phys

MW01534008

Applied Science Review™

Physical Geology

Robin L. Hoffer, DGS
Adjunct Assistant Professor and Lecturer
University of Texas
El Paso

Springhouse Corporation
Springhouse, Pennsylvania

Staff

EXECUTIVE DIRECTOR, EDITORIAL
Stanley Loeb

SENIOR PUBLISHER, TRADE AND TEXTBOOKS
Minnie B. Rose, RN, BSN, MEd

ART DIRECTOR
John Hubbard

ACQUISITIONS EDITOR
Maryann Foley

EDITORS
Diane Labus, David Moreau, Kevin Law, Sarah Thorne

COPY EDITORS
Diane M. Armento, Pamela Wingrod

DESIGNERS
Stephanie Peters (associate art director),
Matie Patterson (senior designer)

ILLUSTRATORS
Philip Ashley, John Gist, Judy Newhouse

MANUFACTURING
Deborah Meiris (director), Anna Brindisi,
Kate Davis, T.A. Landis

EDITORIAL ASSISTANTS
Caroline Lemoine, Louise Quinn, Betsy K. Snyder

Cover: *Gemstone crystals in matrix. Scott Thorn Barrows.*

Printed in the United States of America.
ASR10-010794

Library of Congress Cataloging-in-Publication Data
Hoffer, Robin L.
    Physical Geology / Robin L. Hoffer.
      p. cm. — (Applied science review)
    Includes bibliographical references and index.
    1. Physical geology. I. Title. II. Series.
QE28.2.H63 1995
550—dc20   94-11325
    ISBN 0-87434-607-X   CIP

Contents

Dedication

To all the students who use this book. To my husband Jerry, my daughter Kelly, my granddaughter Alisha, and my mother Lois—for their patience and encouragement. And last, but not least, to my friend Grace for helping me with this book and to my good friend and department chairman, Dr. G. Randy Keller, for suggesting it.

Preface

This book is one in a series designed to help students learn and study scientific concepts and essential information covered in core science subjects. Each book offers a comprehensive overview of a scientific subject as taught at the college or university level and features numerous illustrations and charts to enhance learning and studying. Each chapter includes a list of objectives, a detailed outline covering a course topic, and assorted study activities. A glossary appears at the end of each book; terms that appear in the glossary are highlighted throughout the book in boldface italic type.

Physical Geology provides conceptual and factual information on the various topics covered in most introductory physical geology courses and textbooks and focuses on helping students to understand:

- the chemical and mineral composition of the Earth's interior and exterior surface
- the nature of volcanoes and igneous, sedimentary, and metamorphic rocks
- the dynamic effects of glaciation, weathering, metamorphism, earthquakes, wind, and mass wasting
- the characteristics of deserts, the sea floor, mountain belts, and continental crust
- the theories of sea-floor spreading and plate tectonics
- the importance and origin of geologic resources.

1

Overview of Physical Geology

Objectives

After studying this chapter, the reader should be able to:
• Discuss the basic principles of physical geology.
• Describe how differences in the composition of the earth's internal structure are instrumental in forming surface features.
• Explain the significance of the theory of plate tectonics.
• Describe the relationship between igneous, sedimentary, and metamorphic rocks in the rock cycle.
• Discuss the role of the hydrologic cycle in shaping the earth's surface.
• Explain why the concept of geologic time was a major breakthrough for interpreting the earth's history.

I. The Science of Physical Geology

A. General information
1. **Physical geology** is the study of the materials composing the earth, as well as the processes and forces that continually shape the earth's surface
2. Scientists study geology not only to learn how and why the earth's surface features originated, but also to help locate mineral resources
3. The principles of physical geology help scientists identify potential geologic hazards, such as earthquakes or landslide-prone areas

B. Basic principles of physical geology
1. The theory of uniformitarianism, formulated by Scottish geologist James Hutton (1726-1797), is considered the basic principle of geology
   a. *Uniformitarianism,* based on the notion that the present is the key to the past, assumes that natural processes (such as erosion and sedimentation) occur at a more or less uniform rate; consequently, scientists can study the ancient rock record to learn, among other things, how and when the sedimentary layers formed
   b. Hutton reasoned that gradual changes over long periods of time could explain the earth's features but, given the slow rate of sediment accumulation, the earth would have to be much older than previously thought for the rock layers of the geologic time scale to have been deposited
     (1) Until Hutton's time, the age of the earth was dictated by church doctrine based on Biblical history and genealogies; the exact date of the

earth's creation, determined by Archbishop James Ussher (1581-1665), was believed to be October 22, 4004 B.C.

(2) Today, scientists generally accept that the earth is approximately 4.6 billion years old

c. Geologists now feel that the principle of uniformitarianism is too simplified and, although the laws of nature have remained the same, the rate and intensity of change has varied over geologic time

2. Another principle credited to Hutton is the principle of **cross-cutting** relationships, which is used to interpret the relative ages of events (for example, an igneous intrusion or a fault must be younger than the rocks it intrudes or cuts)

3. Danish anatomist Nicolas Steno (1638-1686) observed sediment transport and deposition during stream flooding in Florence, Italy, and is credited with the following principles

a. The principle of **superposition** states that, in an undisturbed succession of sedimentary rock layers, the oldest layer is on the bottom and the youngest on the top; thus, geologists can interpret the sequence in which the rock layers were deposited

b. The principle of **original horizontality** states that all sediment was originally deposited in layers nearly parallel to the earth's surface; using this information, geologists can tell if rock strata has been moved (tilted and uplifted)

c. The principle of *lateral continuity* states that sediments extend horizontally in all directions until they thin out or terminate against the edge of a basin; thus, geologists can map the extent of a sedimentary layer

4. William Smith (1769-1839) generally is credited with the principle of **fossil succession,** which states that fossil species succeed one another in a definite and recognizable order (a **fossil** is the remains, trace, or imprint of a plant or animal preserved in the earth's crust since prehistoric time)

C. Earth's structure

1. The earth is divided into concentric layers that result from density differences caused by variations in composition, temperature, and pressure of materials (see *Earth's Layers*)

a. The **crust,** the earth's rigid outer layer, is composed of granitic continental crustal material and basaltic oceanic crustal material

b. The **mantle,** the middle and thickest of the concentric layers, is composed principally of peridotite (a dark, dense, igneous rock containing abundant iron and magnesium)

(1) The **lithosphere** is the brittle, solid portion of the earth (about 100 km deep) that encompasses the crust and part of the upper mantle

(2) The **asthenosphere** is the layer of the earth below the lithosphere (also in the upper mantle); because of extreme heat and pressure, the rocks in this layer are weak and begin to bend and flow without breaking

c. The **core,** the innermost layer, is divided into a liquid outer portion and a solid inner portion, both of which are thought to consist of iron-nickel alloys

2. Knowledge of the earth's interior is based, in part, on analysis of seismic data, which involves the compilation of the arrival times of thousands of seismic waves caused by earthquakes at stations around the world; these waves pass through the earth, bending and adjusting their courses to allow for density differences in the earth's layers

Earth's Layers

Structurally, the earth is composed of concentric layers of materials that differ in composition, temperature, and pressure. This cross-section through the earth's interior shows the crust, mantle, inner core, and outer core. Note that the lithosphere includes both crust and upper part of the mantle and the asthenosphere consists of that portion of the upper mantle immediately below the lithosphere.

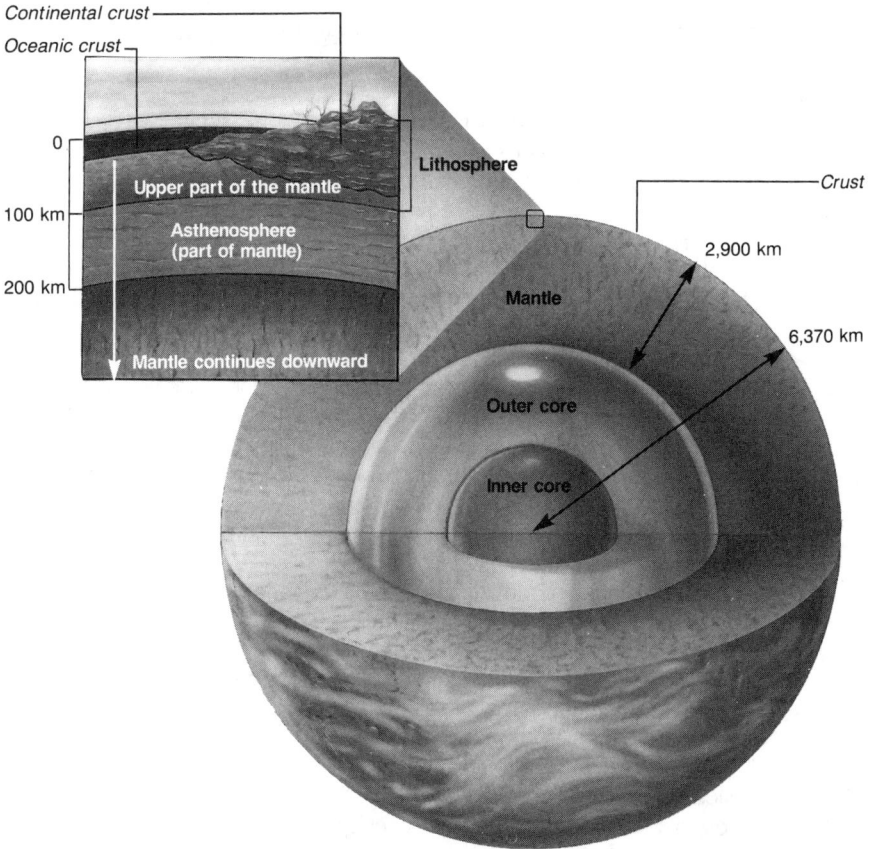

Continental crust

Oceanic crust

0

100 km

200 km

Upper part of the mantle

Asthenosphere (part of mantle)

Mantle continues downward

Lithosphere

Crust

2,900 km

6,370 km

Mantle

Outer core

Inner core

3. Scientists also learn about the earth's interior by studying *meteorites* — metallic or stony objects from interstellar space that have fallen to earth — most of which are believed to be asteroid fragments and consist of solid matter similar to that from which the earth was originally formed

D. Major rock types

1. The rocks of the earth have been classified into three basic groups based on their formation

2. *Igneous rocks* are formed from molten rock material that either solidifies within the earth (such as granitic intrusions) or erupts at the surface (such as basaltic lava flows)

3. *Sedimentary rocks* are formed from the compacted or cemented weathered products of preexisting rocks (for example, sandstone and shale), from the re-

mains of plants and animals (such as coal and some limestone), or from chemical precipitates (such as rock salt)
4. *Metamorphic rocks* are formed when preexisting rocks and minerals are
changed by heat, pressure, or chemically active fluids beneath the earth's surface (for example, gneiss and marble)

II. Interactions Between Earth's Internal and External Layers

A. General information
1. Interactions between the earth's interior and exterior are controlled primarily by
three processes: *plate tectonics,* the *rock cycle,* and the *hydrologic cycle*
2. The driving force behind these processes is heat energy
   a. The external heat source is the sun, which drives the atmosphere and the hydrologic cycle
   b. The internal heat may originate from several sources: residual heat resulting
   from the earth's cooling from a molten state; heat produced by the decay
   of radioactive isotopes; or heat produced by crystallization of the core's liquid portion
3. The mechanism by which earth plates move involves the formation of *convection
   cells,* which develop as hot mantle material rises from the earth's interior,
   moves laterally, then cools, and begins to sink
4. Certain changes on the earth's surface are thought to be directly attributed to the
   interactions of these interior and exterior forces
   a. Mountains have been uplifted by interior forces and worn away by agents of
   erosion (such as running water)
   b. Continents riding atop moving plates have changed size, shape, and position
   c. Ocean basins have closed as continents collided; new ocean basins have
   opened when they split apart
   d. New oceanic crust is produced at midocean ridges (which exist at divergent
   plate margins) and destroyed at subduction zones (which exist at convergent plate margins)

B. Plate tectonics
1. The theory of plate tectonics, which is accepted by most geologists, accounts for
   seemingly unrelated geologic phenomena, including the connection between
   earthquakes, the sea floor, and the origin of different types of igneous rock
2. Scientists believe that plate tectonics has played a major role in the evolution of
   the earth, especially in the distant past when it is thought to have operated at a
   much faster rate than it does today
3. The theory of plate tectonics developed from the hypothesis that the upper portion
   of the earth (the lithosphere) is broken into seven large plates and several
   smaller ones, and that these plates move independently across the earth's surface, sliding on the asthenosphere
4. Most plates are composed of oceanic crust and continental crust (only the Pacific
   plate is entirely oceanic crust material) over a layer of upper mantle material
5. Plate tectonics demonstrates how the forces within the earth are reflected on the
   earth's surface
   a. Folded mountain ranges are found along regions where two continental
   plates have collided (for example, the Appalachian Mountains)

   b. Volcanic eruptions occur where plates are being subducted and are undergo-
      ing partial melting (for example, the Cascade Range volcanoes)
6. Plate tectonics is based on elements of two earlier concepts: continental drift and
   sea-floor spreading
   a. **Continental drift** (proposed in 1912 by German meteorologist Alfred
      Wegener) is based on the assumption that all of the continents were once
      joined together (Wegener called this supercontinent *Pangaea*) and that the
      continents moved or drifted horizontally relative to one another and to the
      ocean basins
      (1) Wegener's evidence included the fit of coastlines, the presence of similar
          rocks and fossils on now separate continents, and paleoclimatic evi-
          dence for polar wandering (the apparent movement in the geologic
          past of the earth's magnetic poles)
      (2) Wegener's theory was dismissed by scientists of his day primarily
          because he had no convincing mechanism by which the continents
          moved through oceanic crust
   b. **Sea-floor spreading** (proposed in the early 1960s by Princeton physicist
      Harry Hess) posits that the continents are not drifting but rather the sea
      floor is spreading
      (1) New sea floor is being formed at midocean ridges when molten rock
          (magma) from the earth's mantle rises upward to the crest of the
          ridge, hardens, and then is pushed aside by the next intrusion
      (2) This new oceanic crust moves away at right angles from both sides of
          the ridge (horizontally), carrying the continents with it
7. Plate boundaries
   a. **Divergent plate boundaries,** or spreading centers, are areas where plates
      are moving apart
      (1) Two types of divergent plate boundaries are recognized (oceanic-
          oceanic and continental-continental), resulting in basaltic lava erup-
          tions and earthquake activity
      (2) New lithosphere forms at these boundaries (such as the mid-Atlantic
          ridge and the rift valleys of Africa)
   b. **Convergent plate boundaries** are areas where two plates collide, resulting
      in deep ocean trenches, volcanic eruptions, and the creation of folded
      mountains; three types of converging plate boundaries are recognized
      (oceanic-oceanic, oceanic-continental, and continental-continental)
   c. **Transform plate boundaries** are areas where two plates are sliding past
      each other, resulting in earthquake activity but no volcanic activity (such
      as the San Andreas fault)

C. The rock cycle
1. The rock cycle is a sequence of events involving the formation, alteration,
   destruction, and reformation of the three major rock groups (igneous, sedimentary, and metamorphic) as a result of magmatism, weathering, erosion, transportation, deposition, lithification, and metamorphism
   a. *Magmatism* is the development and movement of magma and its solidifica-
      tion to igneous rock
   b. **Weathering** is the physical disintegration and chemical decomposition of
      rock

The Rock Cycle

The rock cycle shown here illustrates the relationship between the three types of rock: igneous, metamorphic, and sedimentary. At any time, processes involving heat, pressure, or environmental forces may cause one rock to change into another rock type or material. For example, when magma cools, it becomes igneous rock, which weathering breaks down into sediment. Sediment, in turn, undergoes lithification to become sedimentary rock. Layering of sedimentary rock generates heat and pressure, transforming it into metamorphic rock. Extreme temperatures change the metamorphic rock into magma, thereby completing the cycle. Although the changes are cyclical, the cycle can be interrupted at any step; thus, sedimentary rock may be exposed and weathered, creating new sediment.

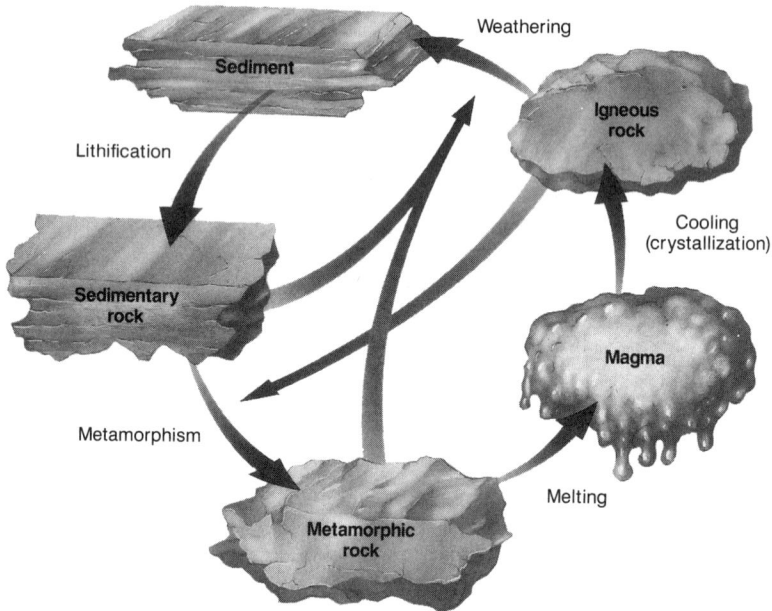

c. **Erosion** is the wearing away of soil and rock by weathering, mass wasting, and the action of streams, glaciers, waves, wind, and underground water

d. *Transportation* involves the movement of sediment by such agents as running water, glaciers, wind, and gravity

e. *Deposition* is the laying down of rock-forming material by any natural agent, such as running water, glaciers, or wind

f. **Lithification** involves the conversion of sediment into solid rock by cementation, compaction, or crystallization

g. **Metamorphism** involves the transformation of rocks either in texture or mineral composition by heat, pressure, or chemically active solutions

2. The rock cycle begins with the formation of magma and its movement toward the earth's surface (see *The Rock Cycle*)

3. When the igneous rock material becomes exposed at the earth's surface — either by direct eruption or, if it solidifies before reaching the surface, by later uplift

and erosion of the overlying rock layers — the rock is subjected to weathering, erosion, or other forces and eventually is transformed into sedimentary rock

4. As the sedimentary rock layers accumulate, the bottom layers begin experiencing increasingly higher temperatures and pressures; in response, textural or mineralogic changes transform the sedimentary rocks into metamorphic rocks
   a. Textural alignment of minerals occur in response to the more confined space
   b. Minerals begin recrystallizing to forms that will accommodate a more confined space
5. With increased temperatures and pressures, metamorphic rocks melt, forming igneous rocks, thereby completing the cycle
6. The rock cycle can be interrupted at any phase and restarted (for example, metamorphic rock can be uplifted, weathered, and changed into sedimentary rock, which can be weathered into loose sediment)
7. Evidence of the rock cycle can be found in various places around the world
   a. New igneous rock material is found at divergent plate boundaries (midocean ridges) and at convergent plate boundaries (volcanic island arcs)
   b. New sedimentary rocks are found along continental margins, across ocean floors, and on continents
   c. New metamorphic rocks are found in areas where sedimentary layers (which may accumulate to great thicknesses on the sea floor adjacent to continents) experience increasingly high temperatures and pressures when the underlying oceanic crust undergoes subduction at a convergent plate margin, or where an igneous intrusive body bakes the enclosing rocks

D. The hydrologic cycle

1. The hydrologic cycle describes the constant circulation of water from the sea, the land, or plants (by evaporation or transpiration) to the atmosphere (where the vapor condenses into clouds) to the land (where it is precipitated in the form of rain or snow) and back to the sea (via rivers and groundwater); see *The Hydrologic Cycle,* page 8
2. The hydrologic cycle is significant because moisture not only aids the chemical breakdown of rocks and minerals in the earth's crust, but also supplies the moisture for running water and glacial ice — important forces in shaping the landscape
3. The hydrologic cycle is essential to human existence; groundwater, the source of most drinking water, is formed when rain or meltwater from snow or glacial ice percolates into the ground (some of the water we drink may have been in the ground for millions of years)

III. Geologic Time

A. General information

1. *Geologic time* marks the period from the formation of the earth to the beginning of human history
2. The ***geologic time scale*** is a means of subdividing time based largely on the relative dating of sedimentary rocks and fossils
3. Beginning in the 18th century, scientists in western Europe noted that rock layers occurred in sequences and that certain fossils and fossil assemblages could

Hydrologic Cycle

The hydrologic cycle describes the continuous circulation of water in the earth–atmosphere system: the most important processes involved are evaporation, transpiration, condensation, precipitation, runoff, and groundwater movement.

Evaporation is the process by which water is transferred from the liquid to the vapor state; the evaporation process is powered by heat from the sun (water is evaporated from oceans, lakes, streams, and soil, and rises into the atmosphere).

Transpiration is the evaporation of water from plant surfaces; most is diffused into the atmosphere through minute pores on leaves.

Condensation, the change from vapor to liquid, occurs when the molecules of water vapor in the atmosphere cool and condense onto microscopic dust particles; this causes clouds to form.

Precipitation occurs when the water droplets coalesce and fall to earth in the form of rain, snow, hail, or sleet.

Runoff is rain or meltwater from ice or snow that is moved downhill by the pull of gravity; runoff flows into streams, lakes, or oceans.

Groundwater, the water that seeps into the ground and is held in openings in rock and soil, moves toward lower levels under the force of gravity and eventually returns to the sea.

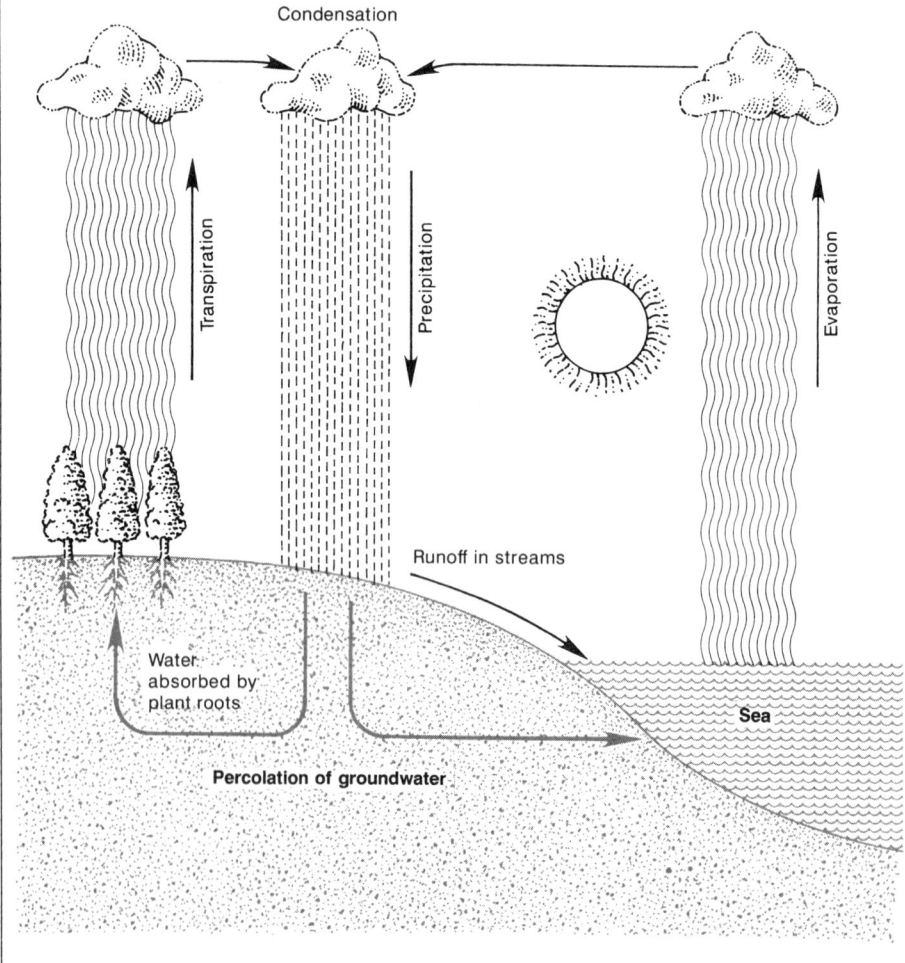

Condensation

Transpiration

Precipitation

Evaporation

Runoff in streams

Water absorbed by plant roots

Sea

Percolation of groundwater

be found in the same rocks; these discoveries served as a basis for the principle of fossil succession

   a. In 1782, French chemist Antoine Lavoisier (1743-1794) demonstrated that different clay quarries near Paris exposed the same sequence of rocks

   b. In the early 19th century, William Smith (1769-1839), an English civil engineer, discovered while surveying and building canals in southern England a method by which the relative age of rock strata could be determined by their fossil content (this was of great economic importance as it saved time and money by knowing the most desirable rock strata through which to build the canals)

   c. In 1810 and 1822, respectively, Georges Cuvier (1769-1832) and Alexandre Brongniart (1770-1847) published maps showing the distribution of rocks around Paris; they noted that a sedimentary bed could be identified by the fossils it contained

B. Development of the geologic time scale

1. The time scale, compiled early in the 19th century, was based primarily on the sequence of fossils found in the rock record (the layers of rock deposited over geologic time is comparable to facts recorded on the pages of a ledger)
2. The time scale also was based on the premise that fossil species succeeded one another in a definite order throughout earth history, and that any time period in the geologic past could be recognized by the fossils left in the rock record
3. Between 1830 and 1842, European geologists began using the principles of superposition and fossil succession as a means of subdividing rock bodies into *systems* (sequences of rocks deposited during a certain period of geologic time) and thus began construction of the geologic column that is the basis for the geologic time scale
4. Systems were named according to rocks laid down during certain intervals of time called *periods,* and these units were then traced to previously unmapped areas (knowledge of rock strata was of economic importance in mining coal, tin, and other ores)
5. Breaks occur in the rock record where rock layers may be missing locally (either removed by erosion or never deposited)
6. The place containing the most complete record of each system was designated the *type section;* this is the location of the originally described sequence of strata that constitute a stratigraphic unit, preferably in an area where the rocks show maximum thickness and are completely exposed
7. Many geologic time periods are named for the locality where they were first described; for example, rocks of the Cambrian Period were first described in Wales (Cambria is the ancient name for Wales)
8. The geologic time scale was based on *relative time,* which is the sequential chronologic ordering of features, fossils, or events (such as igneous intrusion or faulting) relative to one another rather than in terms of years
9. Once the geologic time scale was established, geologists worldwide could identify rocks of the same age by their fossil content
10. With the discovery of radiometric dating (which uses the known decay rate of certain radioactive isotopes), geologists have been able to assign *absolute time* (dates in years) to the subdivisions of the geologic time scale

C. Units of geologic time

1. The *Precambrian time* refers to all geologic time (4.5 billion years to 570 million years ago), and its corresponding rocks, before the beginning of the Cambrian Period
   a. This period has been subdivided into the *Archean Eon* (3,800 to 2,500 million years ago) and the *Proterozoic Eon* (2,500 to 570 million years ago)
   b. An eon is the longest interval of geologic time
2. Some pre-Archean rocks are thought to exist, but no rocks older than approximately 3,960 million years have been found (these were found in the Northwest Territories of Canada)
3. Primordial earth is thought to have been partly or perhaps entirely molten; the early crust may have been thin and composed of *ultramafic* rock (rock with a silica content of less than 45%) and was being continuously destroyed by subduction
4. The Archean and the Proterozoic Eons are followed by the **Phanerozoic Eon**
   a. This eon is further subdivided into (from oldest to youngest) the **Paleozoic Era,** the **Mesozoic Era,** and the **Cenozoic Era**
   b. **Eras,** which encompass shorter spans of time than eons, are characterized by profound changes in life-forms
5. Eras are further subdivided into periods (interval when a specific rock sequence was deposited and characterized by lesser changes in life-forms) and then broken down into **epochs** and finally into **ages**
6. The epochs of the two periods of the Cenozoic Era are the only ones identified in most beginning texts; in such texts, the epochs of the other periods are simply named lower, middle, and upper
7. Time units are simply subdivisions of geologic time and may be designated as early, middle, and late (for example, the late Cambrian period refers to the time when rocks of the upper Cambrian System were deposited)
8. The ages are named for the predominant life form that existed during that span of time; for example, the Cambrian Period is called the Age of Trilobites (these marine arthropods belong to the class Trilobita and are characterized by a three-lobed outer skeleton)

Study Activities

1. Define uniformitarianism, and explain why this concept has been called the key to the past.
2. List and explain three important principles or laws of geology.
3. Sketch the rock cycle, and explain how plate tectonics recycles rock material.
4. Name the major layers of the earth and describe the composition of each.
5. Briefly outline the theory of plate tectonics, and identify the three types of plate boundaries.
6. Explain how the geologic time scale was constructed.

2

Atoms, Elements, and Minerals

Objectives

After studying this chapter, the reader should be able to:
- Name the three components of the atom and describe how each affects the behavior of the atom.
- Explain how atoms gain or lose electrons to become ions.
- List the types of chemical bonds and explain how they affect the physical properties of a mineral.
- Identify the eight most abundant elements in the earth's crust.
- Describe the physical properties of minerals.
- Describe how silicate minerals are classified.
- Explain why the silica tetrahedron is considered the building block of most minerals.

I. The Atom

A. General information
1. All matter is composed of individual particles called **atoms**
2. There are 106 types of atoms, but only 92 occur naturally (the others were created in laboratories) and these are called elements
3. All atoms of the same element share the same chemical properties
4. The size of the atom is measured in Angstrom units (Å), which is a hundred-millionth of a centimeter; most atoms have a diameter of 2 Å
5. All atoms are electrically neutral

B. Atomic structure
1. All atoms consists of three basic particles — protons, neutrons, and electrons
   a. *Protons* carry a positive charge and have the greatest influence on the chemical behavior of the atom; the number of protons determines the *atomic number* of the element
   b. *Neutrons* carry a neutral charge and reside with protons in the nucleus of the atom; **isotopes** are atoms that vary in number of neutrons
   c. The number of protons and neutrons determine the *atomic mass number;* the *atomic weight* is the weight of an average atom of an element given in atomic mass units (amu)
   d. If an element has no isotopes or only one isotope, the atomic mass number and the atomic weight are the same; if an element has more than one isotope, the atomic mass number may vary

      e. *Electrons* have a negative charge and orbit the nucleus

         (1) For every proton or positive charge in the nucleus, there is an electron or negative charge orbiting it

         (2) Electrons build energy levels (shells) in widening circles around the nucleus

            (a) Shell number one contains a maximum of two electrons (with the exception of hydrogen, which has only one); successive shells contain varying numbers of electrons, but the outermost shell needs eight electrons to be stable

            (b) Atoms strive to obtain stability by filling their outermost energy level with a maximum of eight electrons

            (c) Atoms with fewer than eight stable electrons in their outermost shell bond with others to achieve this "magic" number of eight

2. *Bonding,* or the way in which electrons are shared or released by atoms, is accomplished in one of four ways: ionic bonding, covalent bonding, metallic bonding, and van der Waal's bonding

      a. *Ionic bonding* is a linkage formed by transferring or shifting electrons from one atom to another

         (1) Ionically bonded minerals are moderately hard and have high melting points

         (2) Ionically bonded minerals are poor conductors of heat and electricity

         (3) Many ionically bonded minerals dissolve easily in water (for example, halite)

      b. *Covalent bonding,* a linkage formed by sharing electrons, is the strongest type of chemical bond

         (1) Minerals with this type of bonding may be hard (for example, diamond) and have very high melting points

         (2) Covalently bonded minerals do not readily dissolve in water

         (3) Covalently bonded minerals do not conduct electricity

      c. *Metallic bonding,* a linkage formed by overlapping shells of electrons, results in a solid swarm of electrons that are free to move about; this gives several important physical properties to metallic substances

         (1) Metals can conduct electricity and transmit heat

         (2) Metals can be twisted or beaten into thin sheets without breaking and can be drawn into thin wire

         (3) Metals are dark and heavy because electrons can approach each other as closely as possible (increasing their density and thus inhibiting light)

      d. *van der Waal's bonding,* the weakest of the chemical bonds, is maintained by weak residual charges between sheets of atoms

         (1) The atoms making up the sheets are held by covalent bonds and are stronger, whereas the atoms that connect the sheets together are held only by the weaker Van der Waal's forces

         (2) Minerals with this type of bonding, such as the micas, can be split into sheets using only a fingernail

      e. Most minerals exhibit more than one kind of bonding

3. Atoms have certain electrical characteristics

      a. *Ions* are atoms that have gained or lost an electron; the process by which they do so is called *ionization*

(1) *Cations* are atoms that have lost electrons; because electrons have a negative (−) charge, for each one lost the atom gains one plus (+) charge

(2) *Anions* are atoms that have gained electrons and, consequently, additional negative charges

(3) Ions and atoms behave as if they were spheres of different sizes packed together to form a geometric pattern

(4) The number of positive or negative charges on an ion is called *valence;* this is shown by the number of + or − charges on the chemical symbol (for example, an oxygen atom that has gained two electrons would be shown as O^{-2})

b. *Complex ions* or *radicals* are those in which a cation and an anion bond so strongly that they behave as a single unit during further ionic activity

(1) Complex anions may consist of more than one kind of atom; for example, $(SiO_4)^{-4}$ has one silica atom and four oxygen atoms, whereas $(CO_3)^{-2}$ has one carbon atom and three oxygen atoms

(2) Complex ions bind with other ions to satisfy any unused electrical charges; an example of this would be calcium (Ca^{+2}) combining with the carbonate radical $(CO_3)^{-2}$ to form calcite $(CaCO_3)$

(3) The size of an ion compared with an atom decreases when electrons are given up and increases when electrons are added; ions of similar size and charge can substitute for one another in a *crystal lattice,* a geometric form in which the atoms or ions are arranged in patterns of equally spaced rows and planes

(4) The atoms of a particular crystal are of a specific size and will tolerate substitution of atoms of other elements only if they are of similar size and charge

(5) Because the behavior of atomic structures is more sensitive to the anion, minerals are classified according to their anions

II. Elements

A. General information

1. Elements are substances consisting of only one kind of atom
2. Native elements, of which there are about 20, occur in the pure or native form and are not combined with other elements (for example, gold [Au])
3. Some elements are not stable enough to exist singly and therefore combine with other atoms of the same element to form molecules (for example, oxygen [O_2])

B. Abundance of elements found in minerals

1. The eight most abundant elements — oxygen, silicon, aluminum, iron, calcium, sodium, potassium, and magnesium — constitute over 98% of the earth's crust by weight and volume (see *The Eight Most Abundant Elements in the Earth's Crust,* page 14)
2. Minerals typically are composed of the most abundant elements as well as those elements that can fit into the crystal lattice of another element readily
3. The ionic sizes of the elements determines which ions can easily substitute for one another in the crystal structure; mineral families with this characteristic are said to have *end members*

The Eight Most Abundant Elements in the Earth's Crust

The chart below lists the names of the eight most abundant elements in the earth's crust, the size of that element's atom, the electrical charge on the ion, the ion's size, and the percentage of weight and volume of the ion in the earth's crust. Note that iron has two ionic states.

ELEMENT	ATOM SIZE (Å)	ION	ION SIZE (Å)	WEIGHT %	VOLUME %
Oxygen (O)	0.006	O^{-2}	1.40	46.6	93.8
Silicon (Si)	1.17	Si^{+4}	0.42	27.7	0.9
Aluminum (Al)	1.43	Al^{+3}	0.51	8.1	0.8
Iron (Fe)	1.24	Fe^{+2} Fe^{+3}	0.74 0.64	5.0	0.5
Calcium (Ca)	1.96	Ca^{+2}	0.99	3.6	1.0
Sodium (Na)	1.86	Na^{+1}	0.97	2.8	1.2
Potassium (K)	2.31	K^{+1}	1.35	2.6	1.5
Magnesium (Mg)	1.60	Mg^{+2}	0.66	2.1	0.3

   a. For example, magnesium (Mg) and iron (Fe) can substitute for one another in the olivine family of minerals
   b. If more iron is available for substitution, the mineral fayalite will form
   c. If more magnesium is available, the mineral forsterite will form
4. Because oxygen atoms are so large, oxygen-containing minerals are less dense; therefore, rocks containing these minerals have a lower specific gravity
5. Rocks in the earth's crust contain minerals with an abundance of oxygen atoms, such as silicates, because oxygen makes up 90% of the earth's crust by volume; for this reason, crustal rocks have a low specific gravity (2.7) when compared with the average specific gravity (5.5) calculated for the earth as a whole

III. Minerals

A. General information
1. Scientists have identified about 2,000 mineral species, of which 100 are common but only about 20 are truly abundant
2. Minerals can consist of one element (such as gold or silver) or a combination of two or more elements (such as quartz [SiO_2])
3. Minerals are composed primarily of alternate positively and negatively charged units that may be single atoms, ions, or radicals
4. Because the two most abundant elements in the earth's crust are oxygen and silicon, these silicon-oxygen compounds form the most abundant group of minerals — known as the silicates; other mineral groups include the oxides, carbonates, sulfides, sulfates, and chlorides
5. Although oxygen is the most abundant element, not all minerals are oxides; this is because oxygen readily combines with other common elements, forming complex anions (such as carbonates and sulfates)

Mohs' Hardness Scale

Hardness, or the ability to resist scratching, is one of the physical properties used to describe minerals. Mohs' hardness scale, named after the early 19th century mineralogist Friedrich Mohs, is based on 10 minerals of increasing (although not entirely progressive) hardness. Minerals of equal hardness can scratch each other, whereas those of greater hardness can scratch those with a lower hardness rating. These minerals are listed (softest to hardest) below. The common objects in the far right column also can be used to determine the relative hardness of a mineral.

MINERAL	RELATIVE HARDNESS	COMPARATIVE HARDNESS
Talc	1	
Gypsum	2	Fingernail (2.5)
Calcite	3	Penny (3)
Fluorite	4	
Apatite	5	Knife or glass (5.5)
Orthoclase	6	
Quartz	7	
Topaz	8	
Corundum	9	
Diamond	10	

B. Properties of minerals

1. Minerals are naturally occurring, inorganic, homogeneous, and crystalline solids with certain physical properties and a chemical composition that varies within definite limits
   a. Because minerals occur naturally, synthetic materials are not categorized as minerals
   b. Because minerals are inorganic, they do not involve organic life or its products
   c. They have a uniform chemical composition
   d. Because their regular, systematic arrangement of ions achieve electrical neutrality, minerals are called *crystalline* substances
   e. All minerals have a defined chemical composition that permits only ions of a specific size to substitute into its crystal structure without destroying it
2. Internal structure, or bond strength, determines the physical properties of the mineral (rather than chemical composition); for example, diamond and graphite have the same chemical composition (carbon) but vastly different physical properties because of the way the atoms are bonded
3. Other physical properties include color, hardness, streak, crystal form, cleavage, fracture, specific gravity, luster, magnetism, taste, smell, and feel
   a. *Color* refers to the wavelengths of light absorbed by the mineral; in individual minerals, color may vary
   b. *Hardness* refers to the resistance of a mineral to scratching (see *Mohs' Hardness Scale*)
   c. *Streak* refers to the color of a mineral when powdered; this is one of the most reliable ways of identifying metallic minerals

d. *Crystal form* refers to the internal order of the crystal structure reflected on the surface of the mineral; all minerals belong to one of six crystal systems (see *Crystal Systems*)

e. **Cleavage** — the way a mineral breaks to produce flat, smooth planes — is determined by the directions of weakness between atoms within the crystal (can be one, two, three, four, or six directions)

f. *Fracture* refers to the uneven breaking of a mineral when there are no directions of weakness within the crystal (in other words, all atoms are held by the same strength in all directions); some types of fracture include *hackly* (breakage on jagged surfaces), uneven, *conchoidal* (shell-like), and splintery

g. *Specific gravity* refers to the weight of a mineral compared with an equal volume of water (for example, if you put one quart of lead on one side of a balance scale, you would have to put eleven quarts of water on the other side; therefore, the specific gravity of lead is 11)

(1) Feldspar minerals generally have a low specific gravity because they are composed of large and less dense oxygen atoms

(2) Minerals with a high atomic number generally have a higher specific gravity than those with a low atomic number because they have more mass (such as gold, which has an atomic number of 79 and a specific gravity of 19.3)

h. *Luster* refers to the way light is reflected from the mineral; the types of luster include metallic, nonmetallic, vitreous, silky, resinous, pearly, waxy, dull, and earthy

i. Magnetism, taste, smell, and feel are some of the other physical properties that are useful in identifying specific minerals

IV. Silicate Mineral Group

A. General information

1. Most minerals in the earth's crust are silicates

2. In silicate minerals, every silicon atom in the crystal is bonded tetrahedrally to four oxygen atoms, forming the complex anion (SiO_4); this is called a *silica tetrahedron* (for an illustration, see *Silica Tetrahedron,* page 18)

3. The silica tetrahedron is considered the basic building block of the earth's crust

4. The silica tetrahedron forms a pyramid-like structure with the large oxygen atoms at the four corners and the small silicon atom in the center of the structure

5. Each of the four oxygen atoms carries two negative charges, and the lone silicon atom carries four positive charges; therefore, after bonding, there is still one unsatisfied charge on each oxygen ion (for a total of four negative charges on the complex anion as a whole)

6. The remaining charge on each oxygen atom is satisfied either by bonding with one or more of the other abundant elements or by bonding with another silicon ion in an adjacent tetrahedron (one oxygen atom shared between two silicon ions)

B. Classification of the silicate minerals

1. Silicate minerals are classified according to the structural arrangement of the silicate anions (see *Common Silicate Structures,* page 19)

Crystal Systems

A crystal lattice is a geometric form in which the atoms are arranged in patterns of equally spaced rows and planes. The atoms of a particular crystal are a specific size and tolerate substitution only if the atoms of other elements are of similar size and charge. When this orderly internal arrangement is seen on the external surface of the solid, it produces what is called a crystal.

Because only a limited number of configurations meet all of these criteria, minerals are confined to one of six crystal systems. The six crystal systems are listed according to symmetry—from most symmetrical to least symmetrical. Common minerals belonging to each system are included.

CRYSTAL SYSTEM	MINERAL	CRYSTAL SHAPES
Isometric	Galena	
Tetragonal	Zircon	
Hexagonal	Quartz	
Orthorhombic	Staurolite	
Monoclinic	Gypsum	
Triclinic	Feldspars	

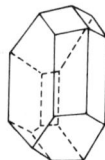

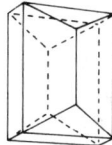

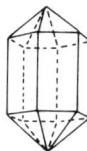

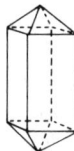

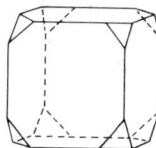

Silica Tetrahedron

The two most common elements in the earth's crust, silicon and oxygen, bond to form a pyramid-like structure, the silica tetrahedron. Four large oxygen ions occupy the corners of the tetrahedron with a small silicon ion at the center of the tetrahedron, equidistant from each oxygen (see the illustration on the left). In reality, the ions touch each other (see illustration on the right).

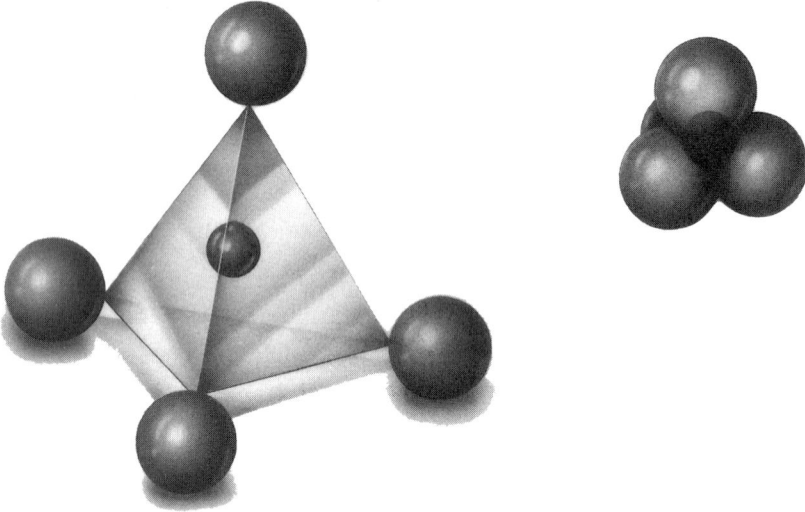

---

2. *Polymerization* is the process whereby two or more tetrahedrons join to form an even larger complex anion
3. Silica tetrahedrons join together at corners, never along edges or faces

V. Rock-Forming Minerals

A. General information
1. Most of the common rock-forming minerals are formed from molten magma
2. Feldspars are the most abundant rock-forming minerals

B. Description of rock-forming minerals
1. The feldspars are divided into two main types: plagioclase feldspars and potassium feldspars
   a. The *plagioclase feldspars,* which are an example of a *continuous mineral series* (one whose composition varies with the substitution of calcium for sodium as the magma cools), range in composition from $CaAl_2Si_2O_8$ at higher temperatures to $NaAlSi_3O_8$ at lower temperatures (a more complete discussion of these feldspars will follow in Chapter 3, Igneous Rocks and Volcanoes)
   b. The *potassium feldspars,* such as $KAlSi_3O_8$, are not as complex as the plagioclase feldspars in that they do not form a continuous mineral series and their chemical formulas do not change

Common Silicate Structures

All 20 common rock-forming minerals are silicates. Silicate structure (listed below) determines the physical properties (cleavage, crystal shape, and so forth) of the minerals. The illustrations in this chart show how adjacent silicon-oxygen tetrahedrons share oxygen ions to form isolated, single-chained, double-chained, sheet, and three-dimensional framework structures. Examples are included of minerals or mineral groups with the silicate structure.

SILICATE STRUCTURE	ILLUSTRATION	EXAMPLE
Isolated silicate		Olivine group
Single chain		Pyroxene group
Double chain		Amphibole group
Sheet silicate		Mica group, clay group
Framework silicate		Quartz, feldspar group

(1) Several potassium feldspars, including orthoclase and microcline, share the same chemical formula but have a different crystal structure

(2) Minerals with this characteristic are called *polymorphs*

2. *Olivine,* $(Mg, Fe)_2SiO_4$, is a ferromagnesium silicate with a single tetrahedron, in which iron and magnesium can substitute for each other in the chemical formula

3. *Pyroxene* is a family of ferromagnesium silicates with a single chain of silica tetrahedrons, the most common one being *augite*

4. *Amphibole* is a family of ferromagnesium silicates with a double chain of silica tetrahedrons, the most common one being hornblende

5. *Mica* is a family of silicate minerals characterized by sheet structures, the most common being *biotite* and *muscovite* (the clay minerals are another important group of minerals that have this sheet structure)

6. *Quartz* is a silicate mineral with a three-dimensional framework that is formed by adjacent tetrahedrons sharing all four oxygen ions

---

Study Activities

1. Select five common silicate minerals and choose which physical properties (hardness, color, and the like) you would use to identify each one. Note that sometimes a mineral can be identified using only two or three physical properties.

2. Sketch a silica tetrahedron, labeling all atoms. Determine the charges on each oxygen atom.

3. Name the types of chemical bonds.

4. Outline the classification of the silicate minerals, and explain how each structure affects cleavage in the mineral.

5. Describe the difference between crystal and crystalline.

6. List the eight most abundant elements in the earth's crust, and give the electrical charge for its ionic form.

3

Igneous Rocks and Volcanoes

Objectives

After studying this chapter, the reader should be able to:
- Explain where magma originates and how it reaches the earth's surface.
- Name the three basic types of magma and describe how they differ in composition.
- Describe igneous rocks and their formation.
- Explain the factors that control the texture and composition of igneous rocks.
- Discuss how Bowen's reaction series derives all magma types from a single parent by differentiation.
- Describe the role that plate tectonics plays in the formation of igneous rocks.
- Compare and contrast the three types of volcanic cones.

I. Magma

A. General information

1. *Magma* is molten or partially molten rock material plus dissolved gases; if magma reaches the earth's surface before solidifying, it is called *lava*
2. Magmas are formed when temperatures within the earth exceed the melting point of the minerals in the rock; water dissolved in magma lowers the melting temperature of minerals, whereas increased pressure raises the melting temperature
3. Magmas form from rocks that melt in the upper mantle or the lower crust, or from a mixture of material from both areas
4. Because magma is less dense than the surrounding solid rock (commonly called *country rock*), it begins rising toward the earth's surface
5. Magmas commonly are classified into three broad categories according to its composition: felsic, intermediate, and mafic; the most common rocks to solidify from these magmas are granites, andesites, and basalts, respectively
6. Magmas are generally siliceous in composition because they form from the melting or partial melting of rocks in which silicates are the main minerals
7. Magmas generally are formed in association with plate tectonic margins
   a. Basaltic (mafic) magmas erupt along divergent plate boundaries or spreading centers (such as midocean ridge systems) and are derived from the partial melting of the underlying mantle; some basaltic magma erupts at hot spots or **mantle plumes,** narrow upwellings of hot material within the

mantle that account for some igneous activity not associated with plate boundaries (for example, the Hawaiian Islands)
  b. Granitic (felsic) and andesitic (intermediate) magma is produced at converging plate margins (such as the circum-Pacific belt) and are thought to result from partial melting of the descending oceanic slab, lower crust, or both

B. Formation of magma

  1. Magma is thought to originate in the earth's upper mantle at depths of 50 to 200 km
  2. Minerals within a rock melt at different temperatures (for example, minerals in a basalt have higher melting temperatures than those in a granite)
  3. When the temperature begins to rise, not all the rock melts at once; the minerals with the lower melting point will melt first
  4. Magma forms only from the mineral components that melt, so the resulting magma may be of an entirely different composition than the parent rock
  5. Several heat sources may be responsible for magma generation
      a. Heat energy released from the decay of radioactive elements can melt rock;
      b. Inherent heat, left over from when the earth was molten, also can melt rock; the *geothermal gradient,* or amount the temperature increases with depth into the earth, is about 25° C/km, or 2.5° C/100 m
  6. The rocks in the earth's interior are prevented from melting by intense pressures; however, when these rocks are brought up from the mantle by convection and the pressure is relieved, they begin to melt
  7. Magma reaches the surface of the earth in several ways
      a. Magma can move upward by *assimilation,* whereby the overlying rock is digested and incorporated into the rising magma
      b. Magma can move upward by *stoping,* whereby the overlying rock is fractured and detached and the blocks are engulfed as the magma moves up
      c. Magma can move upward by following zones of weakness, such as faults and fractures

C. Composition of magma

  1. The composition of a magma depends on the parent rock from which it was formed and the degree to which the parent rock melted
  2. Because rocks of the earth's lithosphere are composed mainly of silicate minerals, the resulting magmas are a mixture of molten silicate material
  3. The most abundant component of igneous rocks (which are generated from magma) is silica dioxide (SiO_2); other important components are oxides of aluminum (Al_2O_3), calcium (CaO), magnesium (MgO), iron (FeO and Fe_2O_3), sodium (Na_2O), and potassium (K_2O)
  4. The *viscosity,* or resistance to flow, is determined by the magma's temperature, percentage of silica, and amount of dissolved gas
      a. The temperature at which magmas solidify can vary from 1,200° C for basaltic magma to 700° C for felsic magma; the hotter the magma, the more fluid the lava flow, and the lower the viscosity
      b. The amount of silica in a magma, which ranges from 45% to 75%, has the greatest influence on the viscosity; this is because the silica tetrahedrons form small structures in the molten material even before crystallization be-

gins, and *polymerization* (the linkage of silica tetrahedrons) advances at a more rapid rate in magmas with a higher silica content

c. The amount of dissolved gases also influences the viscosity of the magma by keeping the magma fluid (dissolved water vapor lowers the melting point of minerals); over 90% of the gases emitted from volcanoes consist of water vapor and carbon dioxide (some of the other gases include hydrogen, sulfur, sulfur dioxide, and hydrochloric acid)

5. The chemical composition of magma determines which minerals will form; the percentage of that mineral, along with other minerals that crystallized at the same time, determines which igneous rock will form

6. Magma is classified into three main types: felsic, intermediate, and mafic

a. *Felsic* (*fel* is an abbreviation for feldspar and *sic* for silica) magmas are rich in silica (65% or more); the remaining percentage (25% to 35%) consists of oxides of aluminum, sodium, and potassium with minor amounts of calcium, magnesium, and iron; rhyolite is the most common extrusive felsic rock

b. *Intermediate* magmas have a composition that falls between felsic and mafic; these magmas contain relatively equal amounts of *ferromagnesium* minerals (minerals rich in iron and magnesium) and sodium- and calcium-rich feldspar; andesite is the most common intermediate rock

c. *Mafic* (*ma* is an abbreviation for magnesium and *fic* for ferric) magmas are considered silica-poor (52% or less); the remaining percentage consists of oxides of aluminum, calcium, magnesium, and iron; basalts are the most common mafic rocks

7. Felsic (granitic) and intermediate (andesitic) composition magmas are produced at converging plate margins, where they form by processes of differentiation, partial melting, and magma mixing

a. **Differentiation** is a process wherein early formed minerals separate from the magma, thereby gradually changing the magma's composition from mafic to felsic

b. Partial melting of basaltic crust could produce minor amounts of a felsic magma, but is more likely to produce magma of an intermediate composition

c. Some felsic magma could form from the melting of silica-rich sediments that are subducted on a descending oceanic plate (these sediments have accumulated over millions of years on the moving oceanic plate)

d. Partial melting of the lower continental crust could form a highly felsic magma and is thought by many geologists to be the probable source for most granitic magmas

e. Intermediate magma, which is less viscous than felsic magma, probably is able to move rapidly through the crust to the surface where it forms andesitic volcanoes; felsic or granitic magma wells up more slowly, crystallizes, and forms **plutonic rock**

f. *Magmatic underplating* posits that mafic magmas pool under the continental crust and act like a hot plate, partially melting the overlying crust and forming granitic magmas

8. Mafic (basaltic) magma is generated by partial melting of upwelling mantle rock along spreading centers (places where the crust of the earth is being pulled

Bowen's Reaction Series

Bowen's reaction series, named for the early 20th century geologist N.L. Bowen, is the sequence in which minerals crystallize from a cooling magma. As temperatures decrease, crystallization takes two distinct but concurrent paths: a discontinuous path and a continuous path. In the *discontinuous path,* minerals undergo a complete change at discrete temperatures, beginning with olivine at the highest temperature, then changing to pyroxene, amphibole, and, finally, biotite at the coolest temperature. In the *continuous path,* changes occur gradually and affect only one mineral—plagioclase feldspar; the feldspar changes from calcic to sodic with decreasing temperatures. The result of these two paths is potassium-rich feldspar, muscovite, and quartz.

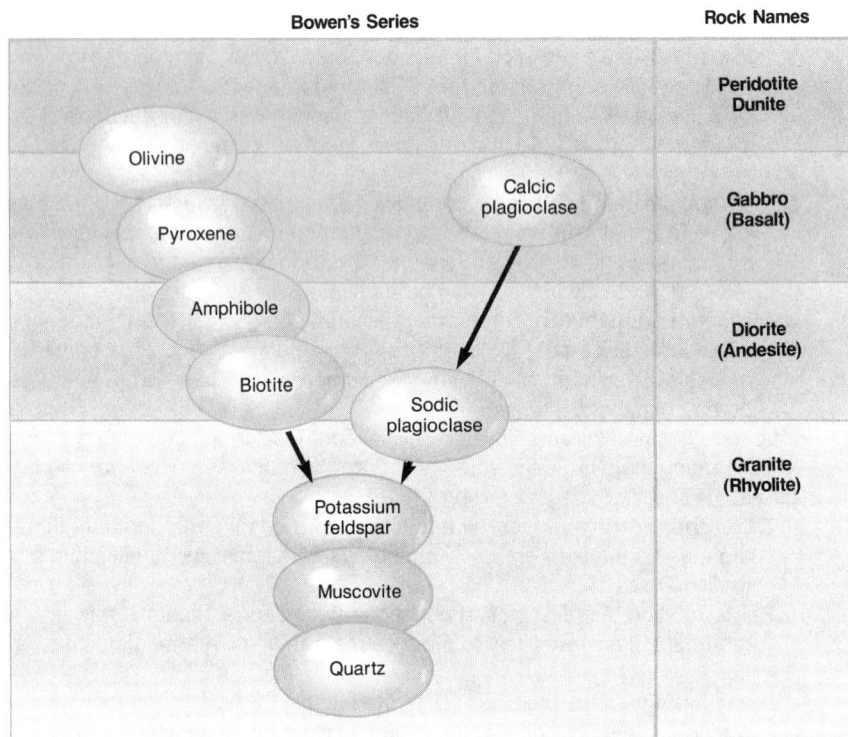

apart by forces associated with plate tectonics) and at hot spots or mantle plumes where upwelling magma is not associated with a plate boundary

D. Consolidation of magma

1. N.L. Bowen discovered that, in cooling silicate melts, some already crystallized minerals can react with the remaining melt; two kinds of reactions exist: discontinuous, in which old minerals react to form new ones, and continuous, whereby a mineral continually changes composition by substituting sodium for calcium in the same crystal structure

2. Bowen demonstrated his hypothesis (now called **Bowen's reaction series**) by describing the sequence of events in a cooling magma of basaltic composition (see *Bowen's Reaction Series*)

    a. With falling temperature, the series has two distinct mineral sequences of crystallization: a *discontinuous series* of olivine to pyroxene to amphibole to biotite, concurrently forming with a *continuous series* of calcic to sodic plagioclase feldspars; these in turn are followed by the sequence of potassium feldspars, muscovite, and quartz

    b. Differentiation, or separation of newly formed minerals, can result in rocks with a different composition than that of the original (for example, felsic magmas, or magma of any composition, could be derived from mafic magmas by differentiation)

    c. Minerals that crystallize at the same temperature occur together in the same rock

    d. Because the composition of the magma changes before the last rocks are crystallized, minerals that form early do not occur in rocks that form later (for example, olivine is not found with quartz)

3. Solubility, rather than melting point, determines the conditions and concentrations at which a mineral will crystallize

4. The composition of a magma is influenced by the partial or total melting of the parent rock

5. The composition of a magma also is changed by assimilating (engulfing and dissolving) the country rock through which it moves (this rock could be of any composition)

II. Igneous Rocks

A. General information

1. *Igneous rocks* form when molten magma cools and crystallizes

2. Igneous *intrusive,* or plutonic, rocks form from magma that solidifies before reaching the earth's surface

3. Igneous intrusive bodies (such as batholiths, stocks, dikes, and sills) are named according to their size and shape in relation to the country rock

4. Igneous *extrusive,* or *volcanic*, rocks form from magma that solidifies after reaching the surface

5. Extrusive igneous rocks can erupt violently and form *pyroclastic* material or produce gently flowing lava

6. Igneous rocks are classified by their composition (mineral content) and texture (mineral or grain size)

B. Intrusive igneous bodies

1. Intrusive igneous rocks are those that apparently intruded preexisting rock as a liquid

2. Intrusive igneous rocks are seen at the surface only after they have been uplifted or the overlying country rock has been eroded away

3. Intrusive igneous rocks are classified by their shape and size and by whether they are discordant or concordant

    a. *Discordant* rock bodies cut across the bedding of the rocks they intrude; they are called batholiths, stocks, or dikes

        (1) *Batholiths,* the largest of the igneous intrusive rock bodies (with an exposed area of greater than 100 km^2), generally have a granitic compo-

Common Igneous Rocks

The chart below groups common igneous rocks by mineral composition and by texture, based on the appearance of mineral grains. The minerals shown are those most commonly found in these rocks, although actual percentages may vary. An asterisk indicates minor amounts.

| | GRANITIC (felsic) | ANDESITIC (intermediate) | BASALTIC | |
			(mafic)	(ultramafic)
Phaneritic (coarse-grained)	Granite	Diorite	Gabbro	Peridotite
Aphanitic (fine-grained)	Rhyolite	Andesite	Basalt	None
Mineral composition	Quartz Orthoclase Sodic-rich feldspar *Biotite *Hornblende	Hornblende Sodic- and calcic- rich feldspar *Biotite *Pyroxene	Calcic-rich feldspar Pyroxene *Olivine *Hornblende	Olivine Pyroxene *Calcic-rich feldspar

sition and occur mostly in younger mountain belts (formed along convergent plate boundaries)

(2) *Stocks* are similar to batholiths except they are smaller (with an exposed area of less than 100 km^2); they probably are part of the same igneous mass as the one that forms the batholith

(3) *Dikes* are tabular, sometimes vertical, intrusive bodies that cut across the bedding of the rocks they intrude; they frequently form radial structures that are emplaced when explosive eruptions split the area around volcanoes and can be composed of either felsic or mafic material

b. *Concordant* rock bodies usually parallel or intrude between the layers or bedding; they are called sills or laccoliths

(1) *Sills* are similar to dikes in that they are tabular bodies; instead of cutting across the layers or beds, sills parallel them

(2) A *laccolith* is a lenticular-shaped body of igneous rock intruded between layers of sedimentary rock; laccoliths dome up the overlying rock strata

C. Classification of igneous rocks

1. Igneous rocks commonly are classified according to texture and composition (see *Common Igneous Rocks*)

2. The *texture* of igneous rocks, which depends on the magma's cooling rate, refers to the size, shape, and arrangement of the individual mineral grains within the rock, producing a pattern of interlocking mineral grains

a. The major types of igneous rock textures are phaneritic, aphanitic, glassy, and porphyritic

b. A *phaneritic* texture reflects a slow cooling history (sometimes millions of years); the mineral grains are large enough to be seen and recognized without a microscope

   c. An *aphanitic* texture reflects a more rapid cooling history; the rock contains a mosaic of minerals, but the minerals are too small to be seen without a microscope
   d. A *glassy* texture reflects a cooling rate that occurred so quickly that only embryonic crystals can be seen with a microscope
   e. A *porphyritic* texture reflects two periods of crystallization and results when crystals formed in a slowly cooling magma are suddenly moved into another environment where the remainder of the magma was cooled more rapidly; the matrix or background is composed of fine crystals in which larger crystals, *phenocrysts,* are scattered
3. The composition of igneous rocks falls into four broad categories: felsic, intermediate, mafic, and ultramafic
   a. *Felsic* rocks are silica-rich (65% or more) rocks with a relatively high content of potassium and sodium (potassium feldspar makes up two-thirds of all feldspar in felsic rocks) and a significant amount of quartz; the two most common types of felsic rocks are granites and rhyolites
      (1) *Granites* are the most abundant of the felsic rocks; they are light-colored, coarse-grained intrusive rocks composed primarily of orthoclase, some plagioclase, quartz, and a minor amount of biotite and amphibole (hornblende)
      (2) *Rhyolites* are the fine-grained equivalent of granites; they commonly are porphyritic (containing orthoclase or quartz phenocrysts) and may have a glassy matrix
   b. *Intermediate* rocks contain between 53% and 65% silica, relatively equal proportions of potassium and plagioclase feldspar, and a minimal amount of quartz; diorite and andesite are the two most common types of intermediate rocks
      (1) *Diorites* are light to dark gray, coarse-grained intrusive rocks that contain nearly equal amounts of potassium and plagioclase feldspar; they lack quartz and have a fair abundance of biotite, amphibole (hornblende), and pyroxene (augite)
      (2) *Andesites* are the fine-grained equivalent of diorites and commonly are porphyritic, with an abundance of plagioclase feldspar phenocrysts
   c. *Mafic* rocks contain less than 52% silica and a large amount of calcic plagioclase feldspar (two-thirds of the total feldspar in mafic rocks), and, rarely, quartz; the two most common types of mafic rocks are gabbros and basalts
      (1) *Gabbros* are dark-colored, coarse-grained igneous intrusive rocks composed of large amounts of calcic plagioclase and a significant abundance of pyroxene (augite) and olivine
      (2) *Basalts* are the dark-colored, fine-grained extrusive equivalent of gabbros and are the most widespread of the igneous rocks
   d. *Ultramafic* rocks are composed mainly of olivine and pyroxene, contain less than 45% silica, and no feldspar or quartz; the most common ultramafic rock is *peridotite,* a dark green, coarse-grained igneous intrusive rock that is thought to be the principal rock of the mantle

III. Volcanism and Extrusive Igneous Rocks

A. General information

1. *Volcanism* is the process by which magma and gases are transferred from within the earth to the surface
2. Volcanoes are the only direct evidence of magma within the earth
3. The violence of a volcanic eruption is directly proportional to the viscosity of the magma (which is dependent partly on its silica content) and the behavior of the gases in the magma
4. Water and carbon dioxide are the two most abundant volcanic gases; the expansion of these gases when they try to escape at the time of eruption causes the still-plastic rocks to explode into pyroclastic fragments
5. The type of volcano formed is directly related to the type of material erupted, which can be felsic, intermediate, or basaltic in nature
6. There are three major types of volcanic cones: shield, composite, and cinder
7. Volcanoes occur in a fairly definite pattern along plate margins (oceanic ridges, rift valleys, and the circum-Pacific area) and as volcanic islands formed over hot spots or mantle plumes
8. Gases emitted from volcanic eruptions are believed to be the source of the earth's atmosphere and oceans; volcanism was more widespread and intense in the earth's geologic past because of the presence of abundant radioactive elements

B. Characteristics of eruptions composed of silicic material

1. Silicic magmas have high silica contents and include both felsic (silica content greater than 65%) and intermediate (silica content between 53% and 65%) magmas
2. Silicic magmas are thick, viscous, and relatively cool; consequently, they hardly flow at all and instead form *domes* over their vent areas (steep-sided protrusions of viscous lava squeezed out of a volcano)
3. The high viscosity inhibits the escape of gas, so tremendous pressures build up, creating explosive and violent eruptions and throwing ash, bombs, and blocks into the air (these pyroclasts are present in basaltic-type eruptions as well, but not in such large quantities)
   a. *Volcanic ash* is fine (less than 2 mm in diameter) pyroclastic material that is carried up by the ascending gas and spread over a large area (which may be several kilometers in diameter)
   b. A *volcanic bomb* is a glob (larger than 64 mm in diameter) of molten lava thrown out of the vent while still viscous, becoming streamlined as it passes through the air
   c. A *volcanic block* is a large angular pyroclast (greater than 64 mm in diameter) ejected from a volcano in the solid state; sometimes it is as large as a house
4. *Nueés ardents,* or *glowing avalanches,* form when incandescent ash, gas, and other pyroclastic material rushes down the side of the volcano; when this superheated ash is deposited, it sticks together, forming *welded tuff* (a glass-rich pyroclastic rock lithified by the combined action of the heat retained by particles, the weight of overlying material, and hot gases)

5. *Calderas,* basin-shaped volcanic depressions (the diameter of which typically is many times greater than that of the included vent or vents) can be formed by explosion (material thrown out of the volcano) or collapse (top of the volcano collapses into the area evacuated by the magma)

C. Characteristics of eruptions composed of basaltic material
1. Basaltic volcanic eruptions are those in which the silica content is 50% or less, and the silica is combined in the feldspar and ferromagnesium minerals; there is no free silica in the form of quartz
2. Basaltic eruptions produce extremely fluid lava flows that can cover large areas and form great thicknesses; some reach speeds up to 40 kilometers per hour (24.8 miles per hour)
3. The two major types of basaltic flows are derived from Hawaiian terms: pahoehoe and aa
   a. *Pahoehoe* flows generally move quickly and produce glassy billows and ropy surfaces
   b. *Aa* flows contain little or no gas and flow slower than pahoehoe flows; as the crust of the moving flow hardens, it breaks up into a jumbled mass of blocks
4. As the gas bubbles rise toward the top of basaltic lava flows, they may become trapped, producing porous or *vesicular* flow tops; rocks with these textures are called *scoria*
5. *Pumice* is formed when the expanding gas forms a froth, resulting in a light, glassy rock
6. As the top of the lava flow cools, polygonal cracks (similar to mud cracks) occur as a result of contraction caused by shrinkage, thus producing *columnar joints*
7. *Lava tubes* are produced when the surface of the lava flow crusts over while lava is still flowing underneath; if the lava continues flowing out from under the crust, a hollow tube remains
8. Basaltic lava also can erupt from fissures or cracks and cover hundreds of square miles; this type of lava is known as *flood basalt*
9. *Pillow structures* form when an elongated blob of lava breaks out of a thin skin of solid basalt during a submarine (underwater) eruption

D. Types of volcanoes
1. *Composite volcanoes* are formed from alternating layers of pyroclastic material and lava flows and can have a felsic or intermediate composition (such as Mt. St. Helens)
2. *Shield volcanoes,* which form broad, gently sloping volcanic cones, are constructed of successive layers of lava generally of basaltic composition (such as Mauna Loa and Kilauea on the island of Hawaii)
3. A *cinder cone* is a steeply sloped (about 30°) volcano constructed of loose pyroclastic material of generally basaltic composition

Study Activities

1. Describe the difference between concordant and discordant igneous rock bodies.
2. List three ways in which magma moves toward the earth's surface.
3. Outline the steps of Bowen's reaction series, and explain how the series can be used to determine the composition of common igneous rocks.
4. Explain how the different types of magma form.
5. Draw the major type of volcanoes, and describe how each develops.
6. Explain how viscosity and gas content determine the violence of a volcanic eruption.
7. List the gases that are given off during a volcanic eruption.

4

Weathering and Soil

Objectives

After studying this chapter, the reader should be able to:
- Differentiate between weathering and erosion.
- Identify the agents of mechanical and chemical weathering.
- Describe how chemical weathering affects the common rock-forming minerals.
- Discuss the relationship between climate and weathering.
- Describe soil formation and how it is affected by climate.
- Explain what soil horizons are and how they develop.
- Differentiate between exfoliation and spheroidal weathering.

I. Weathering

A. General information

1. All rocks are formed in a definite environment under certain temperature and pressure conditions
2. Rocks that are stable when in the earth's interior (such as plutonic and metamorphic rocks) generally become unstable at the earth's surface (as a result of exposure caused by uplift and the erosion of overlying rock layers), where air and water attack them
3. The order of chemical stability of the common rock-forming minerals parallels Bowen's reaction series: the first minerals to crystallize from a magma are the first to weather (see Chapter 3, Igneous Rocks and Volcanoes, for more details about the reaction series)
4. *Weathering* is the physical disintegration and chemical decomposition of rocks exposed at or near the earth's surface; it should not be confused with *erosion,* which is the physical removal of weathered sediment and rock by *mass wasting* (the downslope movement of soil and rock material by gravity) and the action of streams, glaciers, wind, and so forth
5. The two main types of weathering are mechanical weathering and chemical weathering
6. Weathering affects the earth's topography by breaking down the rocks into sediment and making it available for removal by the primary agents of erosion (running water, wind, and the like); this lowers land elevation and wears away mountains
7. Exfoliation and spheroidal weathering both produce rounded surfaces on bodies of rock

8. Climate is an important factor in weathering
   a. Chemical weathering is most effective in subtropical and tropical climates with warm temperatures and abundant rainfall
   b. Mechanical weathering predominates in arid climates, where water is scarce, and in arctic climates, where frost action facilitates rock disintegration
9. Weathering is one of the most important geologic processes; without weathering (especially chemical weathering) there would be no soil formation, and without soil there would be no food supply

B. Mechanical weathering

1. *Mechanical weathering* is the physical breakdown of rocks into smaller fragments without changing rock composition; this process can increase the surface area of the rock and thereby increase the rate of chemical weathering (see *Mechanical Weathering*)
2. The agents of mechanical weathering include frost wedging, frost heaving, the actions of plants and animals, human activity, expansion due to unloading, abrasion during transport, and possibly expansion and cooling of minerals in desert environments (the effects of this last agent are still being debated by geologists)
   a. *Frost wedging* occurs when water entering rock joints and fractures freezes at night, expanding and prying apart the rocks (water expands 9% when it turns to ice)
   b. *Frost heaving* occurs when water freezing in soil causes it to expand, pushing or heaving rocks up to the surface; farmers in New England say their fields grow rocks because, although they remove surface rocks from their fields in the summer, the fields are covered with rocks again by the following spring, as a result of frost heaving
   c. Roots of plants enter cracks in rocks, grow, and force them apart
   d. Animals burrowing or making paths across rock surfaces dislodge and disintegrate rock
   e. *Unloading* occurs when large igneous masses, formed within the earth under high pressure, are exposed at the earth's surface; when this pressure is released, the rocks expand and layers (from less than a centimeter to several meters in thickness) break off in concentric slabs, a process called **exfoliation**
   f. Rocks are further disintegrated by *abrasion,* the result of rock fragments grinding against each other during transport
   g. Salts dissolve in water, wash into basins in arid environments, and then are deposited in rock crevices; when the water evaporates, the resulting salt crystals, which occupy more space than the dissolved form, force apart rocks

C. Chemical weathering

1. Unlike mechanical weathering, **chemical weathering** involves rock decomposition; air and water bring about chemical reactions that turn the rocks and minerals into new ones
2. The agents of chemical weathering are oxidation, hydration, and carbonation
   a. *Oxidation* occurs when minerals react with oxygen in the atmosphere, forming oxides

Mechanical Weathering

Mechanical weathering breaks rocks into progressively smaller fragments, thereby increasing the surface area and making them more vulnerable to chemical weathering. This diagram shows how the surface area of a rock increases as it is broken into smaller and smaller pieces.

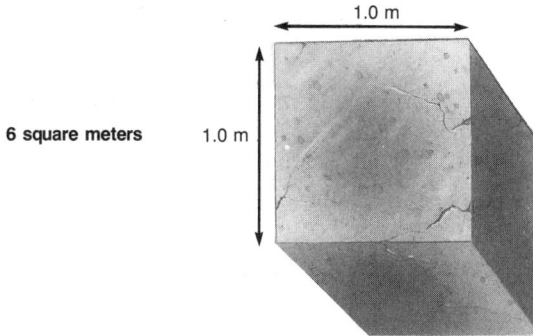

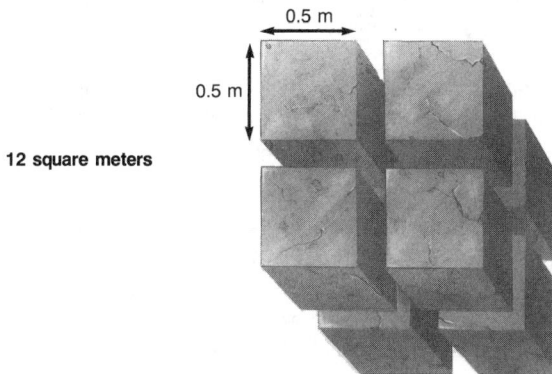

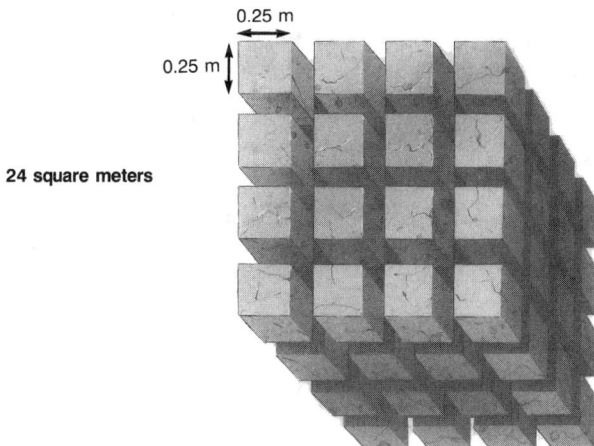

6 square meters — 1.0 m × 1.0 m

12 square meters — 0.5 m × 0.5 m

24 square meters — 0.25 m × 0.25 m

(1) Oxygen, extracted from the air, is incorporated into the mineral's crystal structures, as illustrated in the following chemical reaction:

4 Fe (iron) + 3 O_2 (oxygen) → 2 Fe_2O_3 (iron oxide)

(2) The common rock-forming (iron-bearing) minerals — olivine, pyroxene, amphibole and biotite — are the most affected by the oxidation process and form the common iron oxide minerals hematite (Fe_2O_3) and limonite ($Fe_2O_3 \cdot H_2O$); the cations from these minerals (K+ [potassium], $Ca+^2$ [calcium], Na+ [sodium], and $Mg+^2$ [magnesium]) are released and removed by running water or incorporated into the structure of clay minerals

b. *Hydration* occurs when minerals react with water, incorporating it into their crystal structures and forming new minerals (such as in the formation of limonite)

(1) In *hydrolysis,* part of the water dissociates into hydrogen ions (H^+) and part of it dissociates into hydroxyl ions (OH^-); the H^+ ion replaces the cations in the silicate minerals (K^+, Na^+, Ca^{+2}, and Mg^{+2}), releasing them into solution where they are readily taken up in the formation of clay minerals

(2) Not all the sodium, calcium, and magnesium liberated from minerals is incorporated into the structures of clay minerals; some are carried to streams and eventually to the oceans, where calcium and magnesium form limestone and dolomite

c. *Carbonation* occurs when carbon dioxide (CO_2) combines with rainwater (H_2O) in the atmosphere, forming carbonic acid (H_2CO_3) — a weak acid that dissolves limestones and changes feldspar minerals to clay minerals; carbonic acid can further react with the calcite ($CaCO_3$) in limestones and form water-soluble calcium bicarbonate

3. Although ordinary rain is weakly acidic (contains carbonic acid) and does little harm to the environment, the burning of fossil fuels (coal, oil, and the like) adds additional carbon dioxide to the atmosphere as well as nitrogen and sulfur, which form nitric acid and sulfuric acid; these acids form the environmental problem *acid rain*, which can kill forests and cause marble statues and buildings constructed of limestone (which are dissolved by acids) to weather faster than normal

D. Effects of weathering on common rock-forming minerals

1. The common rock-forming minerals include the feldspar minerals, the ferromagnesium minerals (primarily olivine, pyroxene, amphibole, and biotite), quartz, and calcite

2. Because feldspar is the most abundant mineral in the earth's crust, its weathered products, the clay minerals, also are abundant

3. The feldspar minerals are altered to clay minerals as they react with acidic water; water and carbon dioxide form carbonic acid, which dissociates into the bicarbonate ion (HCO_3^-) as in the following equation:

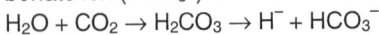

$H_2O + CO_2 \rightarrow H_2CO_3 \rightarrow H^- + HCO_3^-$

II. Soil Formation

A. General information

1. **Regolith,** the unconsolidated rock material that overlies the **bedrock** (solid rock), includes rock debris of all kinds, vegetal accumulations, and soil
2. **Soil** is the upper part of the regolith; it consists of unconsolidated, weathered rock material, water, air, and organic matter that is capable of supporting plant life
   a. *Residual soils* are those that have developed from the weathering of rock directly beneath them
   b. *Transported soils* are those that have developed elsewhere and have been transported and deposited by groundwater, running water, glacial ice, and so forth
3. Soils contain clay minerals, iron oxides, quartz, and organic matter; *humus* is the generally dark, more or less stable part of the organic matter, which may be so well decomposed that its original source cannot be identified
4. Climate is the most important factor in the production of good soils because it controls the amount of chemical weathering, the process necessary for the production of clay minerals (clays form principally from the chemical breakdown of the feldspar minerals)
   a. Temperate climates are warm and wet enough to promote good soils because, unlike tropical climates, the amount of rainfall is not so high that important ions (nutrients) are washed away
   b. Clay minerals also form in tropical soils, but the clays are cation deficient because of leaching by rainwater; if the ground waters have low acidity, even the silica of clay minerals will be leached away, leaving the aluminum oxide mineral, bauxite ($Al_2O_3 \cdot H_2O$), a source of aluminum
5. Clay minerals are essential in the formation of good soils because cations (such as K^+, Na^+, Ca^{+2}, and Mg^{+2}) are easily removed from the sheet structures of clay minerals by plant roots
6. Chemical weathering releases ions that can become part of the clay mineral structure; if clay minerals do not take up these ions, they are removed from the soil and carried to the sea

B. Soil horizons

1. Soil layers are divided into distinctive sublayers called *soil horizons,* which are distinguishable by physical properties, such as color, texture, or chemical composition; they generally are designated by a capital letter (see *Soil Horizons,* page 36)
   a. The *A horizon,* or top layer, is composed of humus and is an area of high biological activity; this also is known as the *zone of leaching* because rain water percolates through it, dissolving the soluble elements and redepositing them in the underlying layer
   b. The *B horizon* lies beneath the A horizon and is called the *zone of accumulation* because material leached from above is redeposited here
   c. The *C horizon* contains the first layer of weathered bedrock; this lies below the B horizon
   d. The area beneath the C horizon is solid bedrock
   e. The *solum,* or true soil, consists of the A and B horizons; these are the zones to which living roots are confined

Soil Horizons

Vertical differences in soil composition, texture, structure, and color divide soil into soil horizons. This diagram shows the A, B, and C horizons in a mature soil. Note that plant roots are limited to the A and B horizons.

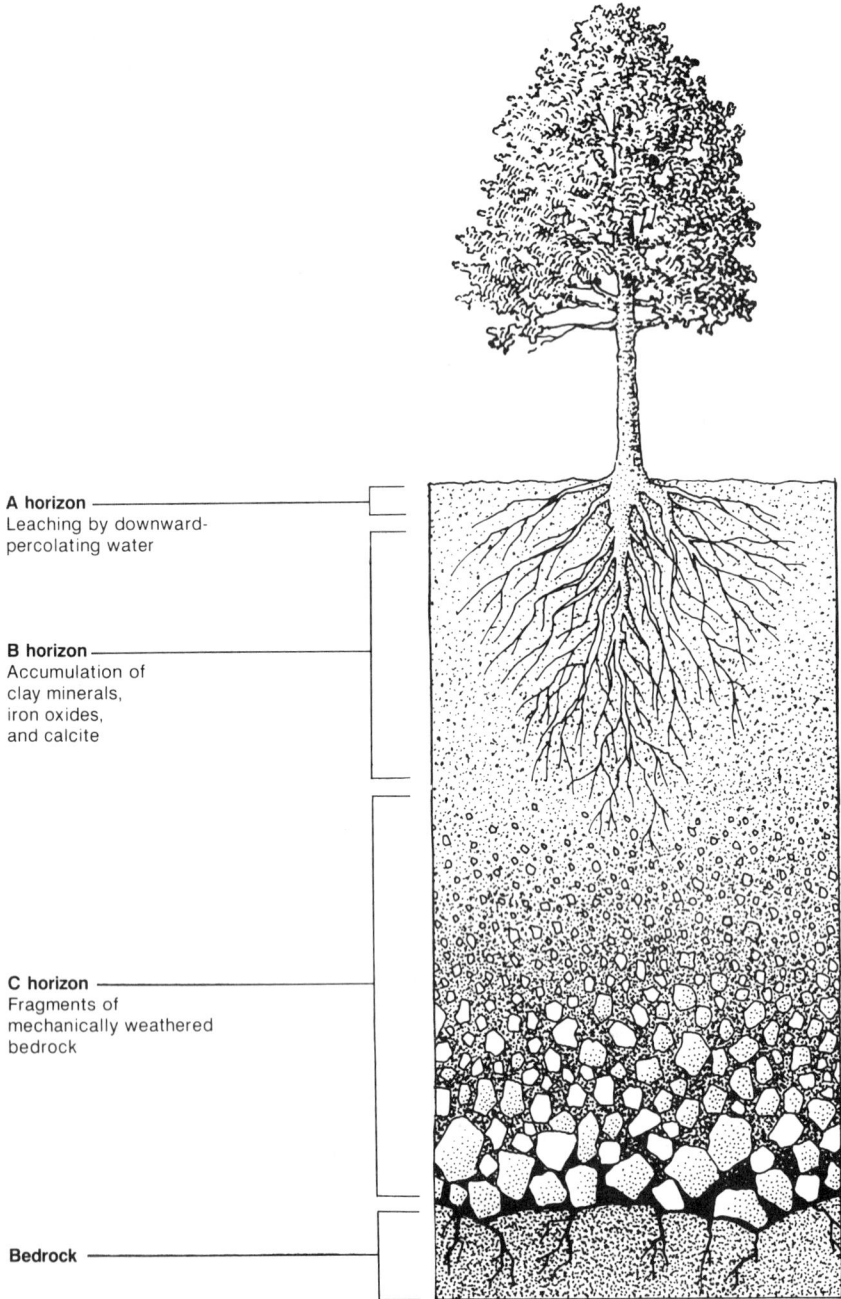

A horizon
Leaching by downward-percolating water

B horizon
Accumulation of
clay minerals,
iron oxides,
and calcite

C horizon
Fragments of
mechanically weathered
bedrock

Bedrock

2. Soils that lack horizons are called *immature* (for example, desert regions may have only the C horizon in the soil because of low rainfall)
3. *Loam,* a common gardening term, refers to a soil composed of equal amounts of sand, silt, and clay, as well as organic matter
4. *Topsoil,* the upper part of the soil, is the same as loam and is more fertile than the underlying *subsoil,* which lacks organic matter and commonly is stony
5. A good soil must be loose, aerated, and well drained and must have an appropriate balance between clay minerals (to hold moisture and provide nutrients) and sand (to help keep soil loose)
6. Three main soil groups are classified on the basis of the mineral content of the A and B horizons
   a. *Pedalfers* are soils containing an abundance of aluminum and iron oxides and clay minerals (*pedon* is the Greek word for soil, and Al and Fe are abbreviations for aluminum and iron, respectively)
      (1) These soils tend to be thick and form in regions where chemical weathering predominates (such as those commonly found in the eastern United States)
      (2) Abundant rainfall promotes leaching, which removes soluble elements and leaves behind insoluble elements, such as aluminum and iron; acids produced by decay of organic material also contribute to the leaching process (H^+ ions replace K^+, Na^+, and so forth)
      (3) Pedalfers develop where the rainfall is not as abundant as in tropical and subtropical climates; thus, some soluble ions remain
   b. *Pedocals* contain an abundance of calcium carbonate (calcite)
      (1) These soils tend to be thin and are characterized by little leaching and scant humus (such as the soils commonly found in the western United States)
      (2) Pedocals form in arid and semiarid regions where the high evaporation rate draws the water in the soil to the surface by subsurface evaporation and *capillary action,* in which water is drawn upward through pores and crevices in the soil
      (3) *Caliche* forms in arid climates and results from the cementing of soil by calcium carbonate and other salt precipitates left behind as water evaporates or as groundwater (which is rich in calcium carbonate) is drawn upward by capillary action and evaporates
   c. *Laterites,* highly leached soil types found in the tropics, are rich in iron and aluminum oxides
      (1) These soils form in regions of high temperatures and abundant rainfall
      (2) Because of the high amount of rainfall, chemical weathering is intense, thereby making the soil nutrient-poor (unable to support good plant growth)
      (3) Tropical rain forests (areas of abundant vegetation and high rainfall) are self-sustaining as long as the forest supplies organic material (dead leaves, twigs, branches, and the like) to the forest floor; if this abundant source of organic nutrients is removed for crop production, the soil quickly becomes nutrient-deficient (because rain washes away any nutrients that the crops have not extracted), and the land becomes barren

Study Activities

1. List four agents that contribute to mechanical weathering.
2. Explain the difference between weathering and erosion.
3. Draw a soil profile, labeling the horizons, the zone of leaching, and the zone of accumulation.
4. Describe the agents of chemical weathering.
5. Describe the type of weathering you would expect to find in tropical, desert, and arctic regions.
6. Explain why soil formation is important to humans.

5

Sediments and Sedimentary Rocks

Objectives

After studying this chapter, the reader should be able to:
- Explain what a sediment is and list the three principal ways sediment originates.
- Identify the agents that transport sediment.
- Compare the common types of environments in which sediment is deposited.
- Describe sedimentary rocks and their formation.
- Explain the classification of sedimentary rocks and list the common types.
- Differentiate between clastic and nonclastic sedimentary rocks.
- Discuss the formation of the common sedimentary structures.
- Define fossils and explain how they can be used to determine ancient depositional environments.

I. Sediment

A. General information

1. *Sediment* is defined as unconsolidated, solid particles created by weathering and erosion of preexisting rock, by chemical precipitation from solution, or by the secretions of organisms in water
2. Sediment is classified and defined according to the size of the individual particles (see *Classification of Sediment Particle Sizes,* page 40)
3. The composition of a sediment is determined by the parent rock and the conditions under which the sediment was generated and transported
4. Sediment is eroded and transported by natural agents, such as running water, wind, and glaciers
5. During transportation, sediment undergoes sorting and rounding
6. Sediment is deposited when the media carrying it (for example, running water) loses energy and can no longer transport it
7. The most distinctive feature of sediment is that it typically is deposited in layers called *strata;* this feature is the one most often retained when sediments are transformed into sedimentary rock
8. Sediment is *lithified,* or converted into sedimentary rock, by compaction, cementation, and recrystallization

B. Sediment origin and composition

1. Clastic sediment originates as accumulations of solid particles, called *clastics,* which are produced from the mechanical weathering of rocks

Classification of Sediment Particle Sizes

Classification of sediment is based on the size of discrete sediment particles. These particles are produced by the breakup of rocks into smaller and smaller fragments. The table below lists the names and sizes for different types of fragments.

NAME OF FRAGMENT	DIAMETER (mm)
Boulder	256 or more
Cobble	64 to 256
Pebble	4 to 64
Granule	2 to 4
Sand	1/16 to 2
Silt	1/256 to 1/16
Clay	Less than 1/256

a. Clastic sediment fragments are classified and defined according to their size, which ranges from (smallest to largest) clay (less than 1/256 mm in diameter), silt, sand, pebble, cobble, and boulder (greater than 256 mm in diameter) (see *Classification of Sediment Particle Sizes*)

b. Clastic sediment fragments may be of any composition (for example, it can be a fragment of an igneous, sedimentary, or metamorphic rock, or a single mineral)

c. It is important to note that the term *clay,* when used to define size, does not necessarily indicate composition and should not be confused with the term *clay mineral,* which is one of a group of silicate minerals with a sheet structure; however, clay minerals generally fall into the clay-sized range

2. Chemical sediment forms from the precipitation of solid mineral crystals from solution; for example, calcite readily dissolves in water (see Chapter 4, Weathering and Soil) and can be precipitated out of water when conditions change according to this chemical reaction:
$CaCO_3$ (calcite) + H_2O + CO_2 (carbon dioxide) $\rightarrow Ca^{+2}$ + 2 HCO_3 (dissolved ions)

3. Organic sediment forms from the secretions of organisms; for example, coralline algae secrete calcite

4. Sediment that originates from rocks directly below them (that have not yet been removed by erosion) is called *residual sediment,* whereas sediment removed from its point of origin and deposited elsewhere is called *transported sediment*

5. Sediment composition is determined by the parent rock, the intensity and type of weathering, and the distance the weathered rock material was transported

a. The weathered parent rock produces a sediment of like composition; for example, a granite composed essentially of the minerals quartz, feldspar, and biotite would mechanically weather to these minerals

b. Sediment composition also is determined by the intensity of weathering; for example, intense chemical weathering of a granite might generate a quartz sand (the feldspars and biotite would decompose to clay minerals, whereas the quartz — the mineral that most resists weathering — would remain relatively untouched)

c. The distance a sediment is transported also affects its composition because the hardness of individual minerals varies; those that are softest would be lost through attrition, whereas those that are hardest would be able to withstand the abrasion encountered during transport (for example, calcite, which has a hardness of 3, would be worn away faster than quartz, which has a hardness of 7)

C. Sediment transport

1. *Sediment transport* involves the physical movement of sediments from one site to another by natural agents (such as running water, wind, glaciers, and gravity)
2. Transportation of sediment from high areas (mountains or plateaus) to low areas (valleys and basins) gradually changes the appearance of the earth's surface because, as the high areas are reduced in elevation, the low areas are built up
3. During transportation, sediment particles are rounded, reduced in size, and sorted
   a. *Rounding* is the process in which the sharp edges and corners of the individual particles are ground away by the impact of other fragments and adjacent rock surfaces; the overall size of the particle also is reduced
   b. The shape of the individual particle is designated as rounded, subrounded, or angular
      (1) *Rounded* fragments are ones whose edges and corners are worn away; large boulders rolling along the bed of a river can be rounded within a mile of their source, whereas smaller fragments have to travel a much greater distance to become rounded
      (2) *Subrounded* fragments are those that show considerable abrasion, having many edges and corners rounded off to smooth curves but with the original form still discernible
      (3) *Angular* fragments are those whose edges and corners remain; this angularity indicates the particles have not been transported very far from their source
   c. Transported sediment undergoes *sorting,* the process by which the sediment particles are selected and separated, according to size and specific gravity, primarily by running water and wind
      (1) Running water transports and sorts more sediment than any other agent
         (a) Water carries sediment in suspension (clay- and silt-sized material) or by rolling and bouncing it along the stream bed (heavier sediment, sand, pebbles, cobbles, and the like)
         (b) Because the finer particles remain in suspension, they are carried farther than the heavier particles and settle out of water together; this process effectively separates the finer, lighter sediment from the coarser, heavier material
      (2) Wind transports and sorts sediment, especially in arid regions where rainfall is scant and vegetation is scarce
         (a) Wind transports the fine sediment suspended in air, while sand and coarser material is bounced or rolled along the ground
         (b) Wind effectively sorts sediment fragments because it carries the finer sediment away, leaving the coarser material behind
      (3) Glaciers transport sediment, either on the surface or embedded in the ice
         (a) Glaciers acquire their sediment load from material that falls onto it from valley walls, from its own abrasion of valley walls, or from the ground over which it moves

(b) Because ice is a more viscous media than water or air (sediment cannot easily settle through ice), the material is very poorly sorted

(c) When glaciers melt, they drop their load; this results in the accumulation of a mixture of different sized fragments

D. Sediment deposition

1. *Deposition* is the laying down of rock-forming material by any natural agent (for example, the mechanical settling of sediment from suspension in water)
2. The most important feature of a sediment is that it is deposited in horizontal layers, with the oldest layer on the bottom; because the layers accumulate over time, they are like pages in a book from which past geologic events can be interpreted (for example, ancient wind-deposited sand dunes preserved in the rock record indicate that when the material was deposited, the area was a desert)
3. Sediment that is transported and deposited is called fluvial, eolian, or glacial sediment
   a. *Fluvial sediment* is transported and deposited by streams; a levee is a form of stream deposit
   b. *Eolian sediment* is transported and deposited by wind; a sand dune is a form of windblown deposit
   c. *Glacial sediment* is transported and deposited by glaciers; a moraine is a form of glacial deposit
   d. Sediment also is moved downslope by gravity
4. Sediment can be deposited in different environments (the characteristics of sediment deposited in each of these environments is discussed in more detail in later chapters); these environments include those on land (lakes, swamps, basins, and so forth), those where streams empty into the oceans (deltas, lagoons, and the like), and those that are deposited in ocean waters of different depths (these marine environments are designated as littoral, neritic, bathyal, or abyssal)
   a. *Lacustrine sediment* is deposited on the floors of lakes
   b. *Paludal sediment* is deposited in swamps
   c. *Basinal sediment* principally consists of clastic material that has been eroded from the highlands that surround the basin (basins are depressions in the earth's surface); when a stream emerges from a narrow canyon onto a plain or a basin floor, its velocity decreases and a fan-shaped pile of sediment called an ***alluvial fan*** is formed
   d. *Deltaic sediment* is deposited where streams flow into bodies of water
   e. *Lagoonal sediment* is deposited in still, *brackish water* (a mixture of fresh and salt water); a lagoon typically exists between a barrier island and the shore or mainland
   f. *Littoral sediment* is deposited in the area between high and low tide
   g. *Neritic sediment* is deposited in the sublittoral or low tide zone and in water up to 200 m deep
   h. *Bathyal sediment* is deposited in marine waters that range from 200 to 1,000 m deep
   i. *Abyssal sediment* is deposited in waters that are more than 1,000 m deep
5. The *environment of deposition* (locality where the sediment is deposited) is marked by certain physical, chemical, and biological conditions; for example, in the brackish water of a lagoonal environment, fine-grained layers of dark shale

containing fossils of oysters and other marine organisms are deposited, and tidal channels filled with coarse sand cut the sediment layers

6. Knowing the characteristics of the various environments on the earth today and the types of organisms that normally live there, geologists can interpret ancient sedimentary environments (this can be useful when prospecting for mineral resources, such as coal or crude oil)

II. Sedimentary Rocks

A. General information

1. A ***sedimentary rock*** is a layered rock formed by lithification, which occurs via compaction, cementation, and recrystallization
   a. ***Compaction*** is the reduction in the volume or thickness of fine-grained sediment resulting from the increased weight of the overlying sediment layers; tighter packing of sediment leads to a reduction in *pore space* (pores are the open spaces between grains)
   b. ***Cementation*** is the process by which sediment particles are converted into rock by the chemical precipitation of mineral material in the pore spaces, binding the grains and forming a solid rock (silica and calcite are common cementing agents)
   c. ***Recrystallization*** is the formation of new mineral grains that essentially are larger than the original grains; when the loose individual minerals forming the sediment recrystallize, they create an arrangement of interlocking crystals (the crystals grow and interfere with each other) that form a solid crystalline rock
2. A sedimentary rock can form from the lithification of any type of sediment, from precipitation from solution, or from the compaction of the remains of plants and animals; these rocks form a thin layer over 75% of the earth's surface and are the site of important resources such as coal, oil, and groundwater
3. A sedimentary rock is classified by its ***texture,*** which refers to the size, shape, and distribution of the particles or crystals that comprise the rock (for example, clastic or nonclastic)
4. Sedimentary rocks formed by the lithification of fragments derived from the mechanical weathering of preexisting rocks have a ***clastic*** texture and are classified as *clastic sedimentary rocks* (for example, sandstone)
5. Sedimentary rocks with a *nonclastic* texture include those formed by chemical precipitation (rock salt and some limestones), biological precipitation (chalk), and accumulation of organic matter (coal)
6. Sedimentary rocks frequently contain fossils that can be used to interpret earth's history; for example, if fossil trilobites are encountered in a sedimentary layer, it is assumed that the rock layer was deposited in an ancient sea during the Paleozoic Era, when these marine arthropods scavenged the sea floor
7. Shale, sandstone, and limestone are the most common types of sedimentary rock; because the feldspars are the most abundant minerals and their main weathering byproduct is shale, the most abundant sedimentary rock is shale

Common Sedimentary Rocks

The chart below groups the common sedimentary rocks into three classes by texture and/or composition:
- Clastic sedimentary rocks are classified according to the size of sediment from which they were formed
- Nonclastic sedimentary rocks, mostly crystalline in texture, are classified by chemical composition
- Organic sedimentary rocks are composed of altered plant material that, when subjected to increased pressures due to burial, is initially converted to peat, then to lignite (a soft brown coal), then to bituminous (a soft black coal), and ultimately—if subjected to the heat and pressure of metamorphism—to anthracite (a hard black coal).

CLASSIFICATION	ROCK NAME	TEXTURE OR COMPOSITION
Clastic	Conglomerate	Rounded gravel
	Breccia	Angular gravel
	Quartzose sandstone	Quartz grains
	Arkosic sandstone	Quartz and potassium feldspar
	Graywacke	Poorly sorted sandstone
	Siltstone/mudstone	Silt and clay
	Shale*/claystone	Clay
Nonclastic (chemically or biochemically precipitated)	Limestone	Calcite ($CaCO_3$)
	Dolostone	Dolomite (Ca, Mg ($CO_3)_2$)
	Rock gypsum	Gypsum ($CaSO_4.2H_2O$)
	Rock salt	Halite (NaCl)
	Chert	Microscopic SiO_2
Organic	Coal	Altered plant remains

*These rocks are called shales if they are laminated.

B. Classification of sedimentary rocks

1. Sedimentary rocks are classified by their origin (compaction and cementation of fragments, chemical precipitation, or accumulation of organic matter) as well as by their texture (clastic or nonclastic)
2. Sedimentary rocks classified by their mode of origin are designated as clastic sedimentary rocks, chemical sedimentary rocks, or organic sedimentary rocks (see *Common Sedimentary Rocks*)
3. *Clastic sedimentary rocks* are composed of rock particles or mineral grains broken from preexisting rock that have become lithified by compaction or cementation of the sediment particles; clastic sedimentary rocks include conglomerate, breccia, sandstone, siltstone, and shale
   a. *Conglomerate* is a sedimentary rock composed of well-rounded cobble and pebble-sized fragments transported by a river or ocean waves and cemented with silica, calcite, or iron oxides
   b. *Breccia* is similar to a conglomerate, but the fragments typically are angular; the particles composing the breccia have not been in transport long enough to become rounded
   c. *Sandstone* is composed of any sand-sized grains or minerals cemented together; quartz typically is the most abundant mineral in sandstone because of its resistance to weathering

(1) *Quartzose sandstone* is composed mainly of the mineral quartz

(2) *Arkosic sandstone* is composed of approximately 25% feldspar with a less than 20% *matrix* (the finer-grained material enclosing the larger grains) of clay, sericite, and chlorite

(3) *Graywacke* is a coarse-grained sandstone consisting of poorly sorted angular to subangular grains of quartz and feldspar, with more than a 15% fine-grained matrix; these sandstones were deposited by turbidity currents

d. *Siltstone,* a massive mudstone in which silt predominates over clay, is *indurated* silt (consolidated by pressure, cementation, or heat) with a texture and composition similar to shale but lacking its fine lamination or fissility (planes along which the rock splits readily)

e. *Shale* — historically, a general class of fine-grained rocks — is a detrital sedimentary rock formed by the compaction of clay, silt, or mud and characterized by a finely laminated, fissile structure

f. *Mudstone,* a blocky, fine-grained sedimentary rock containing approximately equal amounts of silt and clay, is indurated mud that has the texture and composition of shale but lacks its fine lamination

g. *Claystone* is indurated clay that has the texture and composition of shale but lacks its fine laminations

4. *Chemical sedimentary rocks* are formed by the precipitation of minerals from seawater or by the actions of organisms and include limestone, dolostone, rock gypsum, rock salt, chert, chalk, and diatomite

a. *Limestone,* the most abundant nonclastic rock, is composed of the mineral calcite ($CaCO_3$), which can be directly precipitated from seawater or formed by organisms such as lime-secreting algae and reef corals

b. *Dolostone* — composed of the mineral dolomite $CaMg(CaCO_3)_2$ — generally is formed by the replacement of calcite

c. *Chert,* a hard, dense microcrystalline rock consisting chiefly of interlocking quartz crystals, occurs as nodules and layers and may be inorganically or organically precipitated

d. *Rock gypsum,* composed of the mineral gypsum $CaSO_4 \cdot 2H_2O$, is formed in restricted basins (water flows in but cannot flow out) during prolonged periods of saline lake or seawater evaporation

e. *Rock salt* is a chemical precipitate that accumulates in restricted basins during prolonged periods of saline lake or seawater evaporation

f. *Chalk,* a type of limestone, is composed of the accumulated skeletal remains of microscopic marine plants and animals that formed extensive deposits on the ancient sea floor

g. *Diatomite* is a soft, white rock composed of the siliceous accumulated remains of microscopic plants

5. *Organic sedimentary rocks* form from the accumulation and decay of abundant plant material in water that has a low oxygen content (generally a swamp or bog); the decay of plant material continues until all oxygen is consumed, the material becomes buried and compressed by the weight of accumulating sediment, and water or other volatile compounds are driven out

a. *Peat* (the initial stage in coal formation) is a brown, lightweight, unconsolidated to semi-consolidated deposit of plant remains

b. *Lignite* is a soft, brown coal (coal changes from brown to black as the amount of carbon increases) formed when peat is compacted; lignite may still contain visible particles of wood

c. *Subbituminous* and *bituminous* (soft, black coal) coals often exhibit banding (layers of different plant material); this type of coal ignites readily and burns with a smoky flame

d. *Anthracite* (hard coal) is actually a metamorphic rock formed during the regional compression associated with folding; when ignited, this hard, black coal burns with a smokeless flame

C. Sedimentary structures

1. *Sedimentary structures* are features that formed either during or shortly after the sediment was deposited but before it was lithified; some of the common structural features of sedimentary rocks include stratification, cross-bedding, graded bedding, mud cracks, and ripple marks

    a. **Stratification,** the distinctive arrangement of sediment into horizontal layers called *beds* (generally distinguishable from one another and perhaps different rock types), typically is retained when the sediment is hardened into rock

    b. *Cross-bedding* is the deposition of inclined layers of sediment (usually sand) that form a distinct angle to the horizontal *bedding plane* (a nearly flat surface separating two rock layers); cross-bedding is characteristic of wind-blown deposits (such as sand dunes), ocean current deposits (such as ocean floor sand ridges), and deltaic deposits that form at the mouths of rivers

    c. **Graded bedding** occurs when sediment of varying size, shape, and density is deposited in layers that grade from coarse at the bottom to fine at the top; for example, a **turbidity current,** a dense current laden with suspended sediment, moves swiftly down a subaqueous slope and spreads horizontally on the floor of the body of water

    d. *Mud cracks* are polygonal patterns of cracks formed when drying mud shrinks; they are preserved in the geologic record when additional sediment is laid down on top of them, thereby filling the cracks with coarser sediment

    e. *Ripple marks* are a series of small, almost equally spaced ridges of sand or fine sediment formed by wind action or the motion of waves in shallow water

2. Mountain-building events (periods of folding and faulting) can disturb the original horizontal position of sedimentary rock layers; because it is important to know in what order sedimentary layers are deposited when studying earth's history, these sedimentary structures are used to distinguish the top and bottom of a bed

Study Activities

1. Define lithification and describe the three processes that convert loose sediment into sedimentary rock.
2. Describe how coal is formed from compacted plant remains.
3. Explain why shale is the most abundant sedimentary rock.
4. Draw each of the following sedimentary structures: graded bedding, ripple marks, mud cracks, and cross-bedding.
5. Compare the three major agents of transport for sediments (running water, wind, and glaciers) and describe the degree to which each agent sorts the load it carries.
6. List three clastic and three nonclastic sedimentary rocks, and describe the formation of each.

6

Metamorphism and Metamorphic Rocks

Objectives

After studying this chapter, the reader should be able to:
- Define metamorphism.
- List the agents of metamorphism.
- Discuss the types of metamorphism.
- Explain the classification of metamorphic rocks.
- Describe the factors controlling the texture and composition of metamorphic rocks.
- Name the common foliated and nonfoliated metamorphic rocks.
- Identify the metamorphic rocks that form from common igneous and sedimentary rocks.
- Distinguish between metamorphic facies and metamorphic grade.

I. Metamorphism

A. General information
1. All rocks form in a definite environment under certain temperature, pressure, and chemical conditions
2. Rocks in a stable equilibrium in one environment may become unstable in a new environment, and their adjustment to these new conditions may result in the formation of new minerals and rocks; this adjustment or change is called **metamorphism**
3. Changes produced by metamorphism include increased grain (mineral) size, chemical reorganization, and new structural patterns resulting in foliation (the parallel arrangement of minerals)
4. If a rock becomes hot enough to melt, it ceases to be a metamorphic rock and becomes igneous rock; if partial melting occurs, the rock is called a *migmatite*
5. Metamorphic environments include fault zones where rocks may bend and break, contact zones between igneous intrusions and country rock, and deeply buried rocks involved in mountain building
6. *Index minerals* (minerals known to form within a specific temperature and pressure range) allow geologists to recognize low, medium, and high **metamorphic grades;** for example, chlorite, a low-grade index mineral, forms under low temperature and pressure conditions
7. Metamorphic processes may be instantaneous, when rocks are crushed or pulverized in a fault zone, or they may take millions of years, such as when a batholith cools

B. Agents of metamorphism

1. The agents of metamorphism are temperature, pressure, and chemical components, which include fluids, vapors, and gases
2. Factors that influence temperature include extreme depth or subduction along a plate margin and contact with a magma source
   a. The temperature increases with earth depth at a rate of 25° C/km (this is known as the *geothermal gradient*); the deeper a rock layer is buried, the higher the temperature
   b. The geothermal gradient also affects rocks subducted at convergent plate boundaries; these rocks melt as temperatures increase with depth
   c. Rising magma may subject rocks to intense heat
3. Factors that influence pressure include lithostatic pressure and compressive and shear stress produced by pressure
   a. The weight of overlying rock layers produces **lithostatic pressure,** generating forces from all directions
   b. *Compressive stress,* pressure that produces forces from two opposite directions, occurs in mountain-building processes
   c. When forces act parallel to one another but in opposite directions, *shear stress* causes deformation or rupture of the rock layers
4. Chemical changes are influenced by diffusion, transfer of ions by hot fluids, vapors, and gases, dehydration reactions, and decarbonation reactions
   a. *Diffusion,* or the pressure brought about by pore fluids (water that fills pore spaces among rocks), causes ions or atoms to migrate into or out of rock
   b. Hot fluids, as well as vapors and gases, can aid in the transfer of ions and atoms to produce new mineral compounds
   c. *Dehydration reactions* cause the rock or mineral to lose water, as shown by the following chemical reaction:
   $$3\ CaCO_3 \text{ (calcite)} + Ca_2Mg_5Si_8O_{22}(OH)_2 \text{ (tremolite)} + 2\ SiO_2 \text{ (quartz)} \rightarrow$$
   $$5\ CaMgSi_2O_8 \text{ (diopside)} + 3\ CO_2 \text{ (carbon dioxide)} + H_2O \text{ (water)}$$
   d. *Decarbonation reactions* cause the rock or mineral to lose carbon, as shown by the following chemical reaction:
   $$CaCO_3 + SiO_2 \rightarrow CaSiO_3 \text{ (wollastonite)} + CO_2$$

C. Types of metamorphism

1. **Contact metamorphism,** which produces nonfoliated rocks called hornfels, occurs when an igneous intrusive body invades country rock, thereby raising its temperature and infusing it with magmatic fluids; the zone of metamorphism, or *aureole,* around the intrusion can vary from 1 to 10 km and does not involve directed pressure
2. **Regional metamorphism,** which produces such foliated rocks as gneiss and schist, generally occurs over large areas associated with mountain-building events; the preferred mineral orientation and distinct foliation of these rocks, which form deep in the crust and surface after uplift and erosion, are illustrative of the extreme stress that has occurred
3. *Dynamic metamorphism* occurs when strong, unbalanced pressures resulting from **diastrophism** (large-scale deformation caused by mountain-building forces) fold, crumple, and crush rock strata, shear and flatten mineral grains, and create lenses (mylonites) from pebbles

D. Effects of metamorphism

1. Metamorphism results in **recrystallization,** or an increase in grain size, in some rocks and minerals
2. Metamorphism also induces *compositional changes,* which produce a reorganization of the chemical components in the rock, creating minerals that are more in equilibrium with new temperatures and pressures
3. *Textural changes* and new structural patterns develop in response to the directed pressures commonly associated with regional metamorphism; rocks displaying foliated textures may have slaty, schistose, or gneissic textures
   a. A *slaty* texture (sometimes called cleavage) involves the parallel arrangement of fine-grained, platy minerals formed by low-grade metamorphism (for example, slate)
   b. A *schistose* texture displays a strongly foliated, parallel arrangement of platy minerals (micas) or prismatic minerals formed by medium-grade metamorphism (for example, schist)
   c. A *gneissic* texture or structure displays foliation that is more widely spaced, less marked, and often more discontinuous than that of a schistose texture; it is formed by high-grade metamorphism (for example, gneiss)

II. Metamorphic Rocks

A. General information

1. Rocks that have undergone change, excluding melting or weathering, are called **metamorphic rocks;** the fact that they are products of transformation implies they are formed from preexisting rocks
2. In metamorphic rocks, the minerals generally fit together without interlocking (as in the case of igneous rocks) or the introduction of cement (as in the case of sedimentary rocks)
3. Metamorphic rocks are exposed on all continents in regions known as *shields* (areas of exposed ancient rock)
4. Metamorphic rocks make up a large portion of mountain belts' crystalline cores
5. Metamorphic rocks are found in continental interiors overlain by sedimentary rocks

B. Classification of metamorphic rocks

1. Metamorphic rocks are classified by their texture and composition (see *Common Metamorphic Rocks,* for examples of composition)
2. Metamorphic rocks have either a *foliated* or *nonfoliated* texture
   a. Metamorphic rocks with foliated textures include slates, phyllites, schist, and gneiss (commonly produced by regional metamorphism)
      (1) *Slates* are fine-grained metamorphic rocks produced by low temperature and pressure that have a characteristic foliation (*slaty cleavage*), resulting from the parallel arrangement of platy minerals (micas); the parent rock is shale
      (2) *Phyllites* are fine-grained metamorphic rocks in which the mica minerals have increased in size via recrystallization, giving the rock a silky sheen; they form from the further metamorphism of slate
      (3) *Schists* are medium- to coarse-grained metamorphic rocks characterized by strong foliation (schistosity), resulting from the parallel arrange-

Common Metamorphic Rocks

Metamorphic rocks commonly are grouped by texture and composition. The characteristic minerals, parent rocks, and composition of both foliated and nonfoliated rocks are included in the chart below.

ROCK NAME	PARENT ROCK	COMPOSITION
Foliated rock		
Slate	Shale	Clays, micas
Phyllite	Shale, mudrocks	Quartz, micas, chlorite
Schist	Shale, mafic igneous rocks	Micas, chlorite, quartz, talc, hornblende,
Gneiss	Shale, felsic igneous rocks	Quartz, feldspars, hornblende, mica
Migmatite	Felsic igneous rocks	Quartz, feldspars, hornblende, mica
Nonfoliated rock		
Marble	Limestone, dolostone	Calcite, dolomite
Quartzite	Quartz sandstone	Quartz
Hornfels	Shale	Micas, garnets, quartz
Greenstone	Basalt	Chlorite, epidote, hornblende, chlorite
Anthracite	Low-grade coal	Carbon

ment of platy minerals; they form from the further metamorphism of phyllite

(4) *Gneiss* is coarse-grained, banded, or roughly foliated crystalline rock in which the dark and light minerals have segregated into layers and whose composition is similar to that of granite; it forms from the increased intensity of metamorphism of a schist

b. Metamorphic rocks with a nonfoliated texture include marble, hornfels, and quartzite

(1) *Marble,* a coarse-grained metamorphic rock, results from the recrystallization of calcite during contact metamorphism of limestones

(2) *Hornfels,* a fine-grained nonfoliated metamorphic rock, results from the contact metamorphism of shale or marl

(3) *Quartzite,* a coarse-grained nonfoliated metamorphic rock, results from contact metamorphism of quartz sandstone

3. The composition of a metamorphic rock depends on the chemical composition of the parent rock and the intensity of metamorphism

a. *Isochemical* changes involve the rearrangement of elements in the parent rock to form new minerals (for example, calcite recrystallizes to marble)

b. *Metasomatic* changes involve the introduction of new elements (probably brought into the system by fluids emitted from an igneous intrusion) to form new minerals with a chemical composition that differs from the parent rock

C. Metamorphic grade and metamorphic facies

1. *Metamorphic grade* indicates the intensity of metamorphism or the severity of pressure and temperature conditions when metamorphism took place; these grades — designated low, medium, and high — are identified using characteristic mineral associations or facies
2. *Metamorphic facies* are characterized by a particular *mineral assemblage,* a group of minerals that form under restricted temperature and pressure conditions; each type of facies is named after its most distinctive rock or mineral assemblage (for example, the characteristically green mineral chlorite, which forms at a low temperature and pressure, gives rise to the name greenschist facies)
3. Several mineral assemblages that commonly are observed in metamorphosed rock sequences have been carefully studied by geologists, and the temperature and pressure conditions under which these assemblages form are well established
4. *Progressive metamorphism* occurs when a rock is subject to increasing temperature and pressure, resulting in the formation of a sequence of rocks (lowest to highest grade of metamorphism) with a different texture and mineralogy than the parent rock
5. The widely recognized facies resulting from the progressive metamorphism of basaltic rocks are, from lowest to highest grade, blueschist facies, greenschist facies, amphibolite facies, and granulite facies
   a. Rocks of the *blueschist facies* are bluish in color due to the amphibole glaucophane and blue-gray lawsonite; this facies represents low-grade metamorphism
   b. Rocks of the *greenschist facies* are green due to the presence of chlorite, actinolite, and possibly some epidote
   c. Rocks of the *amphibolite facies* contain the dark green mineral hornblende and the plagioclase feldspar andesine
   d. Rocks of the *granulite facies* contain pyroxene and dark red garnet (almandine or pyrope garnet)
6. Rocks produced by increasing grades of metamorphism show the greatest foliation of the metamorphic rock types
7. Metamorphic rock names apply solely to the overall texture of the rock and do not apply strictly to mineral composition; the sequence of foliated metamorphic rocks produced by shale is given below (note that, with increasing grade of metamorphism, the rock texture changes)

Sedimentary rock	→ Low–grade metamorphism	→ Medium–grade metamorphism	→ High–grade metamorphism
Shale	→ Slate	→ Schist	→ Gneiss

8. **Retrograde metamorphism** exposes rocks and minerals to temperatures and pressures lower than those in which they originally formed; for example, dunite, an igneous rock composed mainly of the mineral olivine, is changed into serpentine (a mineral more in equilibrium at lower temperatures and pressures than olivine)

Study Activities

1. Describe the classification of metamorphic rocks.
2. Explain where you would find metamorphic rocks.
3. Illustrate how stress, shear, and lithostatic pressures act on a rock.
4. Compare the three major types of metamorphism, and explain the pressures and temperatures associated with each.
5. List the four most common foliated metamorphic rocks, and describe how each formed.
6. Define metasomatism, and explain how it differs from an isochemical change.
7. Describe which rocks are produced through the progressive metamorphism of shale.

7

Geologic Time

Objectives

After studying this chapter, the reader should be able to:
- Explain relative dating and absolute time.
- Compare the three main types of unconformities.
- Describe how sedimentary rock strata are correlated.
- Differentiate between a fossil and an index fossil, and explain their role in correlating rock strata.
- Discuss how igneous and metamorphic rocks are dated.
- Explain how the decay rates of radioactive elements are used to date rocks.
- Name the major divisions of geologic time.
- Relate when major extinctions of animal and plant groups occurred in geologic time.

I. Relative and Absolute Dating

A. General information
    1. Scottish geologist James Hutton (1726-1797) advanced the principle of *uniformitarianism,* which is based on the assumption that the processes at work on the earth today also operated in the geologic past; this concept can be more simply stated as "the present is the key to the past"
    2. The concept of geologic time is concerned not with hundreds or thousands of years, but millions and billions of years
    3. Early geologists used relative dating to determine the order of events, not how long ago an event happened
    4. Radioactivity was discovered around the turn of the 20th century, and shortly thereafter, the first attempts were made to determine the earth's age using radioactivity (this led to the presently accepted age of 4.7 billion years)
    5. The discovery of radioactive isotope decay rates provided a method for determining the absolute age of rocks; the ages of the rocks of the geologic time scale could now be determined in years

B. Relative dating
    1. *Relative dating,* which uses the sedimentary rock record to determine the chronologic order of geologic events, is based on the following concepts: the principles of original horizontality, superposition, lateral continuity, cross-cutting relationships, inclusions, and fossil succession

     a. The principle of ***original horizontality*** states that all sedimentary rocks were originally laid down in a flat, horizontal position

     b. The principle of ***superposition*** states that the oldest layer is always on the bottom and successive layers are younger toward the top (unless folding or faulting has occurred)

     c. The principle of *lateral continuity* states that sediment extends laterally in all directions until it thins or pinches out

     d. The principle of cross-cutting relationships states that if a rock body, such as a dike, cuts across an existing rock layer, the body into which it intrudes is the older one

     e. The principle of *inclusions* states that, if fragments of one rock are contained in another, the fragment (inclusion) is the older one

     f. The principle of ***fossil succession*** states that life has progressed from simple to more complex forms through time and that fossil assemblages have proceeded each other in a predictable order

2. Relative dating uses not only conformable sequences of sedimentary layers or sequences in which there is no break in the stratigraphic record, but also ***unconformities,*** which are periods where strata are missing due to nondeposition or erosion

3. Three types of unconformities exist (see *Types of Unconformities,* page 56)

     a. ***Angular unconformity*** is an erosional surface that separates layers of tilted sedimentary rocks below from flat-lying sedimentary rock layers above

     b. ***Disconformity*** is an erosional surface between parallel layers of sedimentary rock

     c. ***Nonconformity*** is an erosional surface between an igneous or metamorphic rock and overlying sedimentary rock strata

4. Relative dating also uses ***correlation*** — a means of determining age relationships between rock units from separate areas — to interpret the rock record

5. Correlations can be achieved using similar lithology, lateral continuity, fossil assemblages, and guide or index fossils

     a. Similar lithology (*lithic* means rock) or rock type can be used to correlate rock (for example, a distinctive sandstone unit)

     b. Lateral continuity is the method of physically tracing a rock layer that continues into another area by walking along its extent

     c. Fossil assemblages, or the remains of several different types of organisms found in the same rock layer (those that lived in the same environment at the same time), can be used to correlate rock

     d. Guide or ***index fossils*** clearly establish the age of the rock strata and therefore help correlate rocks

        (1) They are easily identifiable

        (2) They lived only a short span of geologic time

        (3) At the time they lived, they were geographically widespread

C. Absolute dating

1. *Absolute dating* is determined using the decay rate of certain radioactive elements; this method gives the specific time the mineral crystallized from molten magma

     a. *Parent isotopes* are unstable elements that spontaneously break apart or decay to a more stable *daughter isotope* (see *Common Long-Lived Isotope Pairs*, page 57)

Types of Unconformities

Three important types of unconformities—substantial breaks or gaps in the stratigraphic sequence—are possible. An *angular unconformity* (A) separates the surface of an overlying rock from underlying, more steeply sloping rock. In a *disconformity* (B), an erosional surface separates parallel beds above from parallel beds below. In a *nonconformity* (C), an erosional surface separates overlying sedimentary rock from underlying igneous or metamorphic rock.

A

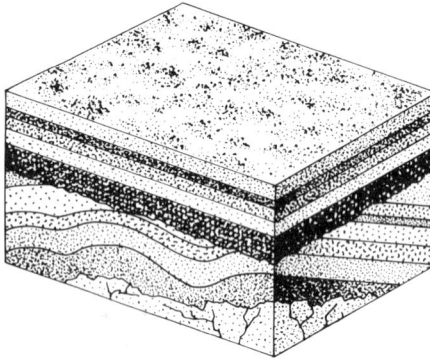

B

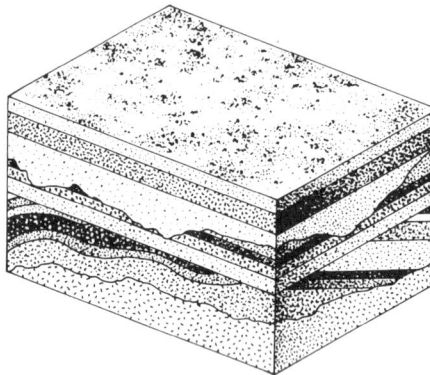

C

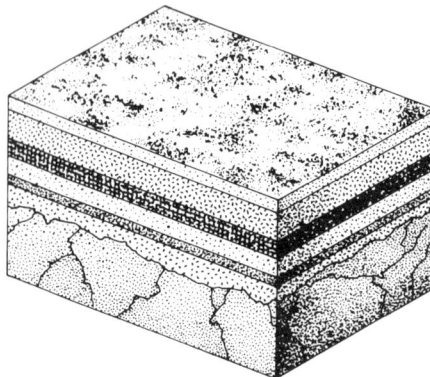

Common Long-Lived Isotope Pairs

The decay rate of radioactive elements—or the time required for half the parent isotope to decay into the more stable daughter product—is a measurable constant that makes absolute dating possible. The most reliable dates are those that use two different radioactive series in the same rock. The chart below lists five parent isotopes and their daughter products used for radiometric dating. Note that carbon dating is used only on organic material (such as bone or wood).

ISOTOPE	DAUGHTER PRODUCT	HALF-LIFE
Uranium 238	Lead 206	4.5 billion years
Uranium 235	Lead 207	710 million years
Rubidium 87	Strontium 87	47 billion years
Thorium 238	Lead 208	14 billion years
Potassium 40	Argon 40	1.3 billion years
Carbon 14	Carbon 12	5,730 years

   b. A *half-life* is the time it takes for one-half the parent isotope to decay into one-half the daughter product; this takes place at a constant, measurable rate
   c. Radioactive decay produces heat as a byproduct of the energy released; geologists and other scientists believe that radioactive decay is the source of the earth's internal heat
2. *Radiocarbon dating* uses the rate of decay of carbon 14 (C14) to carbon 12 (C12), which has a half-life of 5,730 years; this method primarily is used to date organic remains
   a. Carbon 14 forms when nitrogen atoms, bombarded by cosmic rays in the upper atmosphere, lose a proton; this loss of a proton changes the atomic number to that of carbon 14
   b. The *carbon cycle* describes the sequence of C14 and C12 uptake and release in organisms; all living things contain the stable element C12 and, while the organism lives, it ingests C14 in its food supply, and the ratio of C14 to C12 remains constant until the plant or animal dies, at which point the amount of C14 begins decreasing

II. The Geologic Time Scale

A. General information
1. The *geologic time scale* organizes and correlates long periods of time into workable units so that the earth's history can be studied systematically
2. The geologic time scale, which is accepted worldwide, was established between 1830 and 1842 by geologists from Great Britain and Europe using only relative-dating methods
3. Methods for subdividing time are based largely on sedimentary rocks, their fossils, similar lithology, and stratigraphic position

4. Geologic time is divided into *eons* (the longest units of geologic time), *eras* (the next unit of time one magnitude lower than an eon), and so on into *periods, epochs,* and *ages* (see *Geologic Time Scale*)
5. Eras, which include the Paleozoic, Mesozoic, and Cenozoic Eras, are defined by profound worldwide changes in life-forms, whereas periods are characterized by somewhat less profound changes in life-forms
6. Precambrian time encompasses almost 90% of geologic time and extends from the origin of the earth's crust to the beginning of the Phanerozoic Eon (the Cambrian is the first period of the Paleozoic Era, which is the first era of the Phanerozoic Eon)
7. Even though the eras of the Phanerozoic Eon encompass less than 570 million years of geologic time, their rock and fossil record is more complete and studied in greater detail
8. Although all the periods are divided into epochs, only those epochs of the Cenozoic Era are given names (formal names are used for other epochs in the field of paleontology); the epochs of the other periods are simply designated as lower, middle, and upper (these correspond to oldest, middle, and youngest divisions, respectively)

B. Precambrian time
1. The Precambrian encompasses 4 billion years and is divided into two eons: the *Archean Eon* and the *Proterozoic Eon*
   a. The Archean Eon began about 3,800 million years ago (this also is the approximate age of the oldest rock found) and lasted until about 2,500 million years ago
   b. The Proterozoic Eon began at the end of the Archean Eon, 2,500 million years ago, and lasted until the beginning of the Phanerozoic Eon, 570 million years ago
2. The Precambrian had only simple life-forms — algae, bacteria, fungi, and worms — and no organisms with exoskeletons
3. Precambrian-age rocks were subject to mountain-building events, distortion, and metamorphism; any clues in their ancient sedimentary rocks were erased or may be difficult to interpret
4. Rocks that predate the Precambrian time are dated using the igneous and metamorphic rocks found associated with their sediments (for example, geologists may obtain an absolute age from an igneous rock that has intruded into the ancient sediments)

C. The Phanerozoic Eon
1. The *Phanerozoic Eon,* or time of visible life, encompasses geologic time from the beginning of the Paleozoic Era (570 million years ago) to the present
2. The Phanerozoic Eon is subdivided into three eras: Paleozoic, Mesozoic, and Cenozoic (listed from oldest to youngest)
   a. The *Paleozoic Era,* or time of ancient life, lasted about 345 million years and is characterized by life-forms that had preservable hard parts; this time is divided into seven periods (because Paleozoic Era rocks were subjected to less deformation than Precambrian rocks, they are easier to study)
      (1) The *Cambrian* (from Cambria, the Latin name for Wales, where the oldest rocks have been found) is the oldest period of the Paleozoic

Geologic Time Scale

The geologic time scale, which divides the earth's long history into smaller time intervals for purposes of study, was based on relative dating of sedimentary rock strata and the fossil record they contain. The chart depicted here shows the eras, periods, and epochs of the geologic time scale. For each period, the approximate age in millions of years also is shown.

ERA	PERIOD	EPOCH	APPROXIMATE AGE (millions of years)
Cenozoic	Quaternary	Holocene (recent)	
		Pleistocene	
	Tertiary	Pliocene	7
		Miocene	26
		Oligocene	37-38
		Eocene	53-54
		Paleocene	65
Mesozoic	Cretaceous		136
	Jurassic		190-195
	Triassic		225
Paleozoic	Permian		280
	Pennsylvanian		310
	Mississippian		345
	Devonian		395
	Silurian		430-440
	Ordovician		500
	Cambrian		570
Precambrian Time		First multicelled organisms	700
		First single-celled organisms	3,500
		Oldest rocks discovered	4,000
		Age of Meteorites	4,500

Era and the first period to contain extensive and varied forms of life as fossil remains in sedimentary rocks

(2) The *Ordovician* (a name derived from an ancient Celtic tribe, the Ordovices) also was located in Wales; its rocks overlie the Cambrian rocks in this region

(3) The *Silurian* (a name derived from another Celtic tribe, the Silures) overlies the Ordovician and was described from rock strata along the Welsh border

(4) The *Devonian* (named after Devonshire, England) was significant in that its rocks contain evidence of the earliest widespread continental and

desert conditions; the Cambrian, Ordovician, and Silurian Periods were a time of extensive marine deposits (the area apparently was covered by seas)

(5) The *Carboniferous* (this has been divided into the *Mississippian* and the *Pennsylvanian* in the United States) was named for the extensive coal deposits formed during this time

(6) The *Permian* (named after the Perm province of Russia) contains both continental and marine deposits; the most significant fact about this period is the major extinctions of animal and plant groups at the close of this time thought to be the result of the collision of the continents that formed *Pangaea* (the supercontinent)

b. The **Mesozoic Era,** or time of middle life, is characterized by a significant number of modern life-forms and is known as the Age of Reptiles (of which the dinosaurs are the best known); the Mesozoic Era comprises three periods: Triassic, Jurassic, and Cretaceous

(1) The *Triassic* is named after a distinct threefold subdivision of rock strata located and described in Germany (the sequence includes continental red beds, followed by a marine sequence, followed by another layer of continental red beds)

(2) The *Jurassic* is named for rock strata located in the Jura mountains in Switzerland and France, and is famous for the vast dune deposits that make up Zion National Park in Utah

(3) The *Cretaceous,* which means *chalk,* is named for the extensive chalk outcroppings found in France and England; during this period, the seas advanced over large areas of the continents for the last time and the close of this time was marked by mass extinctions of plant and animal groups, including the dinosaurs

c. The **Cenozoic Era,** which means recent life, also is called the Age of Mammals because this is the era in which mammals rose to prominence; the Cenozoic Era is broken into two periods: the Tertiary Period and the Quaternary Period

(1) The *Tertiary* is a name derived from an earlier attempt to divide the earth's history into Primary (now Precambrian) and Secondary (now Paleozoic and Mesozoic Eras), as well as Tertiary and Quaternary, the only two names still in use today

(2) The Tertiary Period is divided into five epochs; from oldest to youngest, these are the Paleocene, Eocene, Oligocene, Miocene, and Pliocene

(3) The *Quaternary* is the last period of the geologic time scale and, as previously mentioned, its name is a remnant of an earlier fourfold subdivision

(4) The Quaternary Period is divided into two epochs: the *Pleistocene Epoch,* which also is called the Ice Age because during this time the earth was covered periodically with great ice sheets, and the *Holocene Epoch,* the last unit of the geologic time scale, is the one in which man lives

Study Activities

1. Draw the three types of unconformities.
2. Explain the difference between absolute and relative time.
3. Describe how C14 is created in the atmosphere, and explain why this method is used to date relatively young events.
4. List three ways rock strata can be correlated.
5. Explain why the geologic time scale was based on relative time.
6. Define uniformitarianism and briefly explain why the concept is so important to geologic thought.
7. List at least three reasons why Precambrian rocks are so difficult to interpret.
8. Explain how we know that seas once covered large areas of the continents.

8

Mass Movement

Objectives

After studying this chapter, the reader should be able to:
• Define mass movement.
• Differentiate between debris and bedrock.
• Discuss the classification of mass movements.
• Explain the rates at which large masses of earth move.
• Name the major categories of downhill movement.
• Distinguish between liquefaction and solifluction, and explain how each occurs.
• Discuss how mass movements are triggered.
• Identify ways of reducing the hazards of mass movement.

I. Classification of Mass Movement

A. General information

1. **Mass movement,** also called mass wasting, is the general term for a variety of processes by which large masses of earth are moved downslope under the influence of gravity
2. Mass movements are classified according to the rate at which the material moves downslope, the type of material moved, and the type of movement
3. The rate at which material moves downslope can be imperceptibly slow (less than 1 cm/year) or very rapid (more than 100 km/hour)
4. The type of material involved can include *debris* (any unconsolidated material at the earth's surface) or **bedrock** (solid rock that underlies soil or other superficial material)
5. The type of downhill movement can include flow, if the material moves as a viscous fluid; slip (slide or slump), if the material moves over the ground; and fall, if the material travels in the air in free fall
6. Mass movements, such as landslides and mudflows, are responsible for loss of life and property damage each year in all parts of the world
7. Geologic investigations, such as mapping, soil and rock analysis, and the construction of slope stability maps to outline areas susceptible to mass movements, can reduce or eliminate the damaging effects of mass wasting

B. Rates of movement

1. Slow mass movement (for example, creep) involves the downslope movement of material at an almost imperceptibly slow rate; this type of mass movement can

transport a much greater volume of material than the more dramatic rapid movements (for example, avalanches and landslides)
2. Rapid mass movement is more dramatic because it is visible and can occur quite suddenly, frequently resulting in loss of life and property damage
3. The rate of mass movement may vary; it may begin as a creep and end up as a landslide

C. Types of mass movement
1. The types of mass movement are classified as flow (which includes creep), slip (which includes slide and slump), or fall; see *Types of Mass Movement*, page 64
2. *Flow* involves extremely slow flowage of material downhill or it may involve debris so saturated with water that it moves rapidly as a fluid; mass movements that involve the flow of material downhill include creep, earthflow, and mudflow
   a. *Creep* involves the slow flowage (may be less than 1 cm/year) of material downslope; evidence of this type of movement can be detected by observing fence posts or gravestones that have moved out of alignment
      (1) *Soil creep* involves the slow movement of topsoil down even a slight slope
      (2) *Rock creep* is the slow movement of rock fragments downslope
   b. *Earthflow* is the downhill movement of debris as a viscous fluid, commonly brought on by heavy rains; the material may remain covered by a blanket of vegetation (involves both slip and flow)
   c. **Solifluction** is a type of flow that predominantly takes place in regions where the ground freezes to a considerable depth (this is called permafrost); as the surface of the permafrost thaws during warm seasons, the upper portion begins moving downslope
   d. *Mudflows* occur when heavy rain or meltwater forms a slurry (thick mixture of soil and water), which typically flows down stream channels
   e. *Liquefaction* is a type of mudflow; this process involves the transformation of loosely packed sediment (usually of clay composition) into a fluid when ground vibrations or added water bring about the collapse of the sediment structure
3. *Slip* involves the movement of a relatively coherent mass (the material remains intact) that moves along one or more well-defined surfaces; slips are further subdivided into slides and slumps
   a. *Slide* involves the movement of material along a plane that generally parallels the slope surface
      (1) *Debris slide* is a rapid downward movement that slides and tumbles forward, forming an irregular or hummocky deposit
      (2) *Rockslide* is the often rapid movement of newly detached segments of bedrock down any planar surface
      (3) *Landslides* — the most rapid type of mass movement — occur when a portion of a hillside, triggered by heavy rains or earthquakes, becomes detached and moves rapidly downslope; rock and debris can move downslope at speeds in excess of hundreds of kilometers per hour
      (4) *Avalanches* originate in high, mountainous regions and descend when the gravity of their snow mass becomes too great for the slope on which they rest

Types of Mass Movement

Mass movement is a general term for gravity-driven processes by which earthen debris moves downslope. These diagrams illustrate the four types of mass movement: flow, slip (slide and slump), and fall.

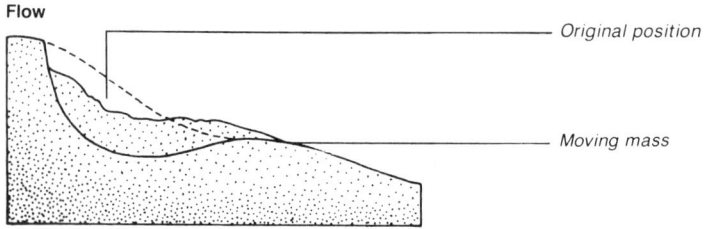

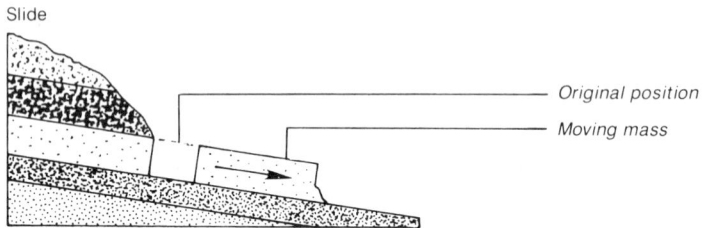

Flow

Original position

Moving mass

Slide

Original position

Moving mass

Slip

Slump

Original position

Slumping mass

Fall

Original position of rock

Falling rock

Waves

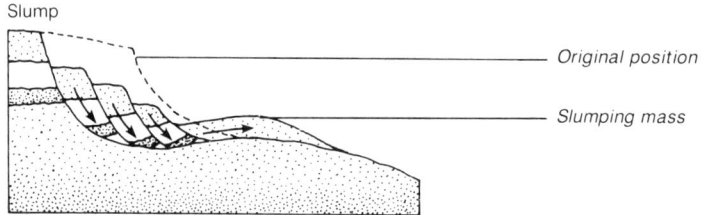

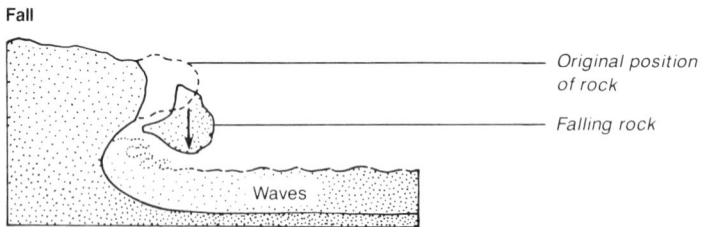

b. *Slump* is the downward slipping of a cohesive mass of rock or unconsolidated material along a curved surface in which the upper part moves downward while the lower part moves forward; slumps tend to develop where strong, resistant rock (such as sandstone) overlies weaker rock (such as shale)

4. *Fall* involves the free-falling movement of newly detached earth debris or bedrock from a cliff or steep slope

a. *Debris fall* is the free fall of unconsolidated earth debris from a cliff

b. *Rockfall* is the free fall of a newly detached segment of bedrock (of any size) from a cliff or a steep slope

II. Factors that Influence Mass Movement

A. General information

1. Gravity is the principal force behind all mass movements
2. Other contributing factors include the addition of water, the steepness of the slope, the amount of vegetation present on the slope, and the type of material involved in the movement
   a. Water, in the form of rain or meltwater, can fill pore spaces between soil particles, thereby decreasing soil cohesiveness
   b. The steepness of the slope on which the material is situated can contribute significantly to mass movements
   c. The amount of vegetation available for holding the material in place may determine the extent and direction of mass movements
   d. The type of material involved in the movement controls the type of mass movement that ensues; for example, angular rock fragments move less slowly than loose soil

B. Factors that control mass movement

1. The gradient (steepness) of the slope on which the material is situated will determine slope stability; generally, the steeper the slope, the more unstable it is
2. Unconsolidated material has an *angle of repose* (the steepest angle that a slope can maintain without collapsing) of between 25° and 40°; the shape and size of the fragments on the slope also contribute to the angle of repose
3. Climate plays an important role in the type of weathering the slope material has undergone (for example, in the tropics where rainfall is abundant, the effects of weathering can extend to depths of several meters); deep weathering produces loose material, which is more easily transported than unweathered rock
4. Water content in the slope material is also a factor; the addition of water, in the form of rain or meltwater, can fill pore spaces between soil particles, causing them to slide past each other
5. Overloading may be the result of human activity, such as dumping, filling, piling up material, or adding water to the slope (water can add weight to a slope whether it is the result of human activity or a heavy rain)

C. Factors that trigger mass movement

1. The addition of water is probably the biggest controlling factor in triggering mass movements; water decreases cohesion, possibly decreases friction between soil particles, and adds weight to the slope
2. Overloading of a slope may occur if the soil becomes saturated by heavy rains; the weight of the water in the soil increases the pull of gravity on the slope, causing it to move downhill
3. Mass movement can be triggered by a river undercutting a slope, by eroding into a bank, or by a bulldozer making a road, thereby causing the slope to become unstable
4. Earthquakes or the vibrations of passing traffic can cause debris to break loose

5. The removal of vegetation also affects slope stability; plant and tree roots tend to hold soil and rock in place

D. Methods used to control mass movement
1. Draining excess moisture from the soil is probably the single most effective way of controlling mass wasting; this can be achieved by adding drainpipes or trenches
2. Retaining walls, with numerous drains, can be to used to hold slopes and prevent further downslope movement of material
3. Keeping native vegetation on slopes allows plant roots to bind soil particles together and hold the soil to bedrock (erosion by running water is particularly damaging after a forest or grass fire)
4. Contour plowing on farm fields with slopes can slow erosion of top soil during rainstorms

---

Study Activities
1. List the ways that mass movements are classified.
2. Compare the different rates of mass movement, and describe how to tell if creep occurs.
3. Describe the climate in which solifluction predominantly occurs.
4. Identify four factors that influence mass movement.
5. Describe the effect of water on slope stability.
6. Define the angle of repose.
7. List three ways downslope movements are triggered.
8. Describe three methods of slowing mass movement.

9

Running Water and Landscape Development

Objectives

After studying this chapter, the reader should be able to:
- Explain how running water originates and how it moves over the earth's surface.
- Distinguish between base level and ultimate base level.
- Identify the factors that influence the velocity and discharge of a stream.
- Explain the processes by which a stream erodes, both vertically and horizontally.
- List the ways a stream transports its load.
- Define what is meant by the competence of a stream.
- Identify the factors that control deposition.
- Explain how natural levees, flood plains, and meanders are formed.
- List the characteristic stages of a developing stream valley.

I. Nature of Running Water

A. General information
1. *Running water* originates as rain or snow meltwater; when it reaches the earth's surface, it flows downslope, pulled by gravity
2. Running water moves as *sheetflow,* the more or less continuous film of water flowing over the land surface after a rain, or as *channel flow,* where the running water is confined to a *channel* (a long trough-like depression)
3. The precipitation reaching the stream channel immediately after a rain is called **runoff** (a runoff also can result from snow meltwater)
4. Runoff in channels commonly is differentiated by volume and size as creek, stream, river, and so forth; however, geologists refer to all runoff confined to channels as *streams,* and the sides of the channel are called *banks*
5. In addition to rainwater and snow meltwater, other sources of water contributing to the stream include **groundwater** or subsurface water
6. Running water flows downslope until it reaches **base level** (the level below which a stream can no longer erode its channel bottom or *stream bed*), which may be another stream, a lake, or a basin, or *ultimate base level* (if the stream empties into the ocean)
7. A stream flows downward from its *headwaters* or source (located at the highest elevation along its course or route) to its *mouth,* the place where the stream enters a larger stream, lake, sea, or basin

8. The *gradient,* or slope down which the stream flows, is defined as the angle between the water surface or channel floor and the horizontal, measured in the direction of flow (the *slope* of the stream)

9. Streams are important for numerous reasons
   a. They are a source of fresh water
   b. They offer a means of transporting goods
   c. They provide water for industry, agriculture, and domestic use
   d. They also are an energy source (electricity can be produced from hydroelectric plants)

B. Stream velocity and discharge

1. *Velocity* is the downhill distance traveled per unit of time; factors that influence the velocity of the stream include its gradient and the shape and roughness of its channel
   a. The primary factor controlling stream velocity is the steepness of its gradient; the steeper the gradient, the greater the velocity
   b. Other factors include the shape of the stream channel (water flows more slowly in a wide channel than in a narrow channel) and the roughness of the stream channel (water flows more slowly over a boulder-covered stream bed than over one with a smooth sandy bottom); as the water travels, it drags along the banks and bed of the channel and the resulting friction slows the water
   c. The velocity of a stream can vary along its course (route) or within a cross-sectional area of the stream itself; for example, velocity is slowest along the banks and bed of its channel and fastest near the middle of its channel because of friction
   d. When a stream goes around a curve or bend, the area of greater velocity is along the outside of the curve (because of centrifugal force), while the area along the inside of the curve experiences lower velocity; thus, the areas along the outside of the curve are eroded, and the areas along the inside of the curve are sites of deposition
   e. The velocity of a stream determines its ability to erode, transport, and deposit its *sediment load* (material carried by the stream); for example, high velocity results in erosion and transportation, whereas low velocity results in deposition

2. *Discharge* is the total volume of water present in a stream as it moves past a certain point (this can be a landform or a particular section of the channel) in a given period of time; it is calculated by multiplying the cross-sectional area of a stream (the width of the stream in feet by the depth of its channel, also in feet) by its velocity (speed is calculated in feet per second)
   a. The discharge of a stream generally increases downstream because water that flows out of the ground into the stream and side streams (called *tributaries*) that empty into the main stream add water to the stream along its route
   b. Discharge can increase during floods (times of high stream flow when water overflows the banks) or decrease if groundwater is lost into the stream bed or by evaporation as it flows through an arid region

C. Stream erosion

1. A stream erodes the sediment and rock over which it flows by solution, abrasion, and hydraulic action
   a. Water in streams can dissolve the rocks over which they flow, removing the more soluble ions (such as potassium, sodium, calcium, and magnesium) and carrying them away in *solution*
   b. *Abrasion* of the stream channel is caused by the impacting, grinding, and scouring action of sediment (ranging in size from sand- to boulder-sized fragments) moving along the stream bed
   c. *Hydraulic action* involves the mechanical loosening and removal of weakly re-sistant material or loose sediment solely by the action of flowing water; it can be caused by a stream impinging against the bank or the outside of a bend or by wave action
   d. Potholes (smooth deep holes or hollows) are formed in the solid rock of the stream bed by the swirling action of water that is filled with coarse sedi-ment
2. Streams deepen and widen their channels by eroding their beds and banks
3. Streams carry away material delivered to them by their tributaries and by mass wasting of their banks

D. Methods of stream transport

1. A stream transports its load — that is, material brought to it by its tributary streams and material produced by its own erosion — as dissolved load, suspended load, and bed load
   a. The *dissolved load* of a stream consists of ions (potassium, calcium, sodium, nitrogen, and so forth) dissolved by the chemical weathering of rock and sediment and brought into the stream by groundwater, runoff, and its own erosion
   b. The *suspended load* of a stream is material that is carried by the stream in suspension, generally silt- and clay-sized particles
   c. *Bed load* is part of a stream's load that is moved on or immediately above the stream bed; it consists of boulders, pebbles, and gravel-sized frag-ments that are transported by sliding and **saltation** (movement along the channel floor by short leaps and bounces)
2. The largest particle a stream can move is defined as its *competence;* competence is related to the stream's velocity — the faster the flow (the more hydraulic lift-ing), the larger the particle the stream can carry
3. The *capacity* of a stream is the total load it can carry; a stream's capacity increases as its discharge increases (because there is more water to carry more sediment), but it may not be able to carry very large fragments (this is classified as low competence)
4. A *graded stream* is one that has attained equilibrium; it has just the right gradient to transport all the material delivered to it by its own erosion, its tributaries, downslope movement, and sheetwash (note that the meaning of a graded stream and the meaning of stream gradient differ greatly)
5. The shape of a stream profile (from its source to its mouth) is roughly concave; at its headwaters, discharge is low and the gradient is steep, whereas down-stream its discharge is greater and its gradient is lower (this is how a stream maintains it equilibrium)

E. Types of stream deposits
1. A stream begins depositing its load as the velocity or discharge diminishes
2. Sites of deposition include braided streams, meandering streams, flood plains, natural levees and deltas, and alluvial fans
    a. A *braided stream* is a stream that divides into an interlacing network of branching and reuniting shallow channels separated by the deposition of numerous *bars* (a mound of sand and gravel that forms in the stream channel or along its banks); a stream tends to be braided when it is heavily loaded with sediment and the discharge is low when compared to the sediment load
    b. A *meandering stream* is a stream with numerous curves or loops; some of their features include meanders, point bars, oxbows, and cutoff meanders
        (1) A *meander* is a bend or loop, produced when the stream swings from side to side in flowing across its flood plain; meanders frequently form in streams approaching base level
        (2) A *point bar* is a sand bar deposited on the inside curve of a meander; sediment is deposited here because the water velocity slows on the inside of meander bends
        (3) An *oxbow* is a closely looping stream meander that has an extreme curvature with only a thin neck of land left between the two parts of the stream
        (4) A *cutoff meander* can form if the stream cuts through the narrow strip of land between adjacent loops of the oxbow, thereby forming an *oxbow lake* (the horseshoe-shaped channel of a former meander) (see *How Oxbow Lakes Form*)
    c. A *flood plain* is that portion of a stream valley adjacent to the channel that consists of sediments deposited during floods; during floods, the stream overflows its banks and spreads out onto the adjacent flat land (velocity and discharge lessen and sediment is deposited)
    d. A *natural levee* is a ridge or embankment of sand and silt built by a stream along both banks in its flood plain; levees are formed predominantly in times of flood when water overflows its normal banks, spreads out, and is forced to deposit the coarser part of its load
    e. A *delta* (named because of its triangular shape, which resembles the Greek letter delta) forms from the accumulated sediment at the mouth of a stream; as the stream empties into the ocean (or another standing body of water), the coarser sediment is deposited near the mouth with progressively finer sediment deposited farther out
    f. An **alluvial fan** forms where streams, which have been confined to a mountain canyon, spread out in a fan-shape on a valley or basin floor; the discharge and velocity of the stream diminishes and sediment is deposited (the coarser, heavier sediment is deposited first, at the mouth of the canyon, and the finer, lighter sediment is carried further out)

How Oxbow Lakes Form

As streams approach a level plain, lateral motion creates a meander. As shown in this illustration, further erosion and sediment deposition result in exaggerated curves. Given time, the stream will cut across the narrow neck separating the two channels and form a new channel, leaving an oxbow lake in the abandoned stream bed.

Meander begins to form

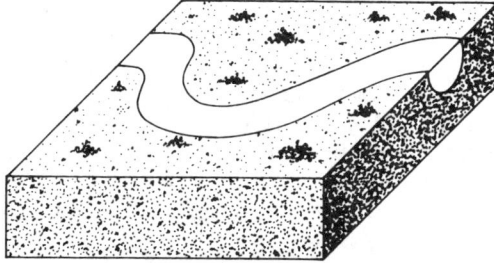

Meander neck narrows

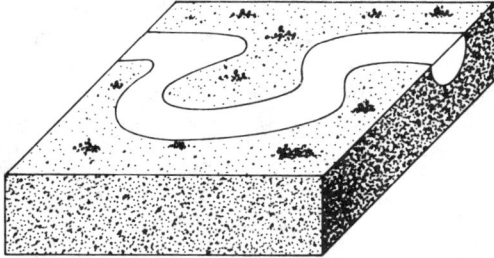

Neck cutoff occurs

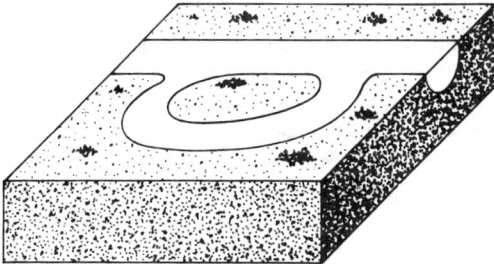

Oxbow lake forms after new channel forms

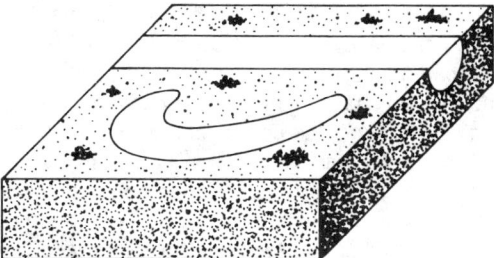

II. Landscape Development

A. General information

1. The landscape or general appearance of the earth's surface is made up of numerous landforms; physical geology is concerned with those landforms created or modified by geologic forces
2. A *landform* is one the many features that together make up the surface of the earth and can include broad features (such as a plain, a plateau, or a mountain) or minor features (such as a hill, a valley, a slope, or a canyon)
3. Running water does more to shape these landforms than any other agent; during heavy rains, a thin film of unchanneled water (sheetwash) flows over the land, rounding the land by wearing away irregularities, and carrying off loose surface material produced by weathering
4. Valleys, the most common landforms on earth, are shaped by stream erosion and undergo progressive stages of development
5. Valleys can intersect other valleys as tributary streams join a main stream, thus forming a drainage system
6. Climate influences the amount and intensity of precipitation received by the stream system
7. A drainage basin is the total area drained by a stream and its tributary streams
8. The development of the drainage pattern of the drainage basin as a whole influences landform erosion
9. Landform development is seen as an instance of the principle of uniformitarianism — that geologic processes operating over long periods of time account for present day landscapes; geologists differ in their opinions on how landforms are eroded to form landscapes

B. Evolution of a stream valley

1. Streams cut their own valleys: deepening them by downcutting, widening them by lateral erosion, lengthening them by headward erosion, and carrying away the sediment that mass wasting and sheetwash delivers to valley floors
   a. *Downcutting* is a valley-deepening process caused by stream bed erosion
   b. *Lateral erosion* is a valley-widening process caused by the undercutting and erosion of stream banks; this results when a stream swings from side to side across its valley floor
   c. *Headward erosion* is the slow, uphill growth of a valley above its original source via gullying, mass wasting, and sheet erosion; the process by which a stream is able to cut back in the direction of its source results in valley lengthening
2. Stream valleys go through several stages of development: extreme youth, youthful, mature, and old age
   a. In the *extreme youth stage* of valley development, as a stream begins downcutting, the valley walls are steep, and rapids and waterfalls are numerous
   b. In the *youthful stage,* as development of the stream valley progresses, mass wasting and sheetwashing erode canyon walls, creating a V-shaped canyon; at this stage, the energy of the stream still is used to downcut its channel, and rapids and waterfalls still are common along its route
   c. During the *mature stage,* lateral erosion starts to widen the valley and a flood plain develops; the mature stage is characterized by meanders, which can occupy the width of the whole valley

       d. During the *old age stage,* the valley becomes wider than the *meander belt* (the zone along a valley floor in which a meandering stream shifts its channel from time to time); meandering causes deposition covering the entire valley, thereby producing a smooth, flat surface

       e. Later uplift of the area *(rejuvenation)* may cause the stream to begin downcutting, producing *stream terraces* (flat, bench-like structures produced by a stream that remains elevated as the stream cuts downward)

       f. *Entrenched meanders* are produced if later uplift of the land causes a meandering stream to start deepening its channel

3. Streams form and flow downhill from their source to their mouth, following the slope of the land, detouring around obstacles, and in general, moving along the path of least resistance; occasionally, however, streams seem to defy nature, cutting across ridges or mountain ranges and forming deep canyons in all types of rock

4. Streams are classified by the way they respond to the landscape on which they formed and are designated as consequent, superimposed (or superposed), or antecedent

       a. A *consequent stream* is one whose course was determined by the surface on which it developed; it conforms to the slope of the land, detouring around hills as would be expected

       b. A *superimposed* or *superposed stream* is one that began its course on horizontal strata, but its channels have deepened and cut across tilted strata

       c. An *antecedent stream* is one that continues to follow a long-established course — regardless of later uplift of the surface over which it developed — by deepening its channel as uplift occurs

C. Stream drainage

1. A stream, such as the Amazon river, consists of a main stream and many smaller tributary streams that flow into it; this makes up a *drainage system*

2. The total area from which the system or systems carry water is known as a **drainage basin** (the drainage basin of the Amazon river covers an area the size of the United States)

3. A *drainage divide* is a ridge or high area that separates one drainage basin from another

4. The term **continental divide** is applied to the primary water-parting area of a continent

5. In the United States, the line separating streams that flow into the Pacific Ocean from those that flow into the Atlantic Ocean is called the Continental Divide; this divide runs along the crest of the Rocky Mountains

6. A *drainage pattern* is the configuration or arrangement of a stream system, which consists of the main stream and all its tributary streams, as they would appear on a map (see *Types of Drainage Patterns,* page 74)

7. Drainage patterns include dendritic, radial, trellis, rectangular, and deranged

       a. In a *dendritic drainage pattern,* the streams form a tree-branching arrangement, indicating that the underlying rocks offer uniform resistance to erosion

       b. In a *radial drainage pattern,* the streams radiate like the spokes of a wheel from a high central area, such as a dome or volcanic cone

       c. In a *trellis drainage pattern,* the main streams are parallel to each other and have right-angle tributaries, which in turn are fed by elongated secondary

Types of Drainage Patterns

These illustrations depict three different types of drainage patterns. The top diagram shows a radial pattern, which can develop on the dome of a volcano. The middle diagram shows a rectangular pattern, which commonly is found on regularly fractured rock. The bottom diagram shows a trellis pattern, typical in regions where tilted layers of resistant rock (such as sandstone) alternate with nonresistant layers (such as shale).

Radial (as viewed on a map)

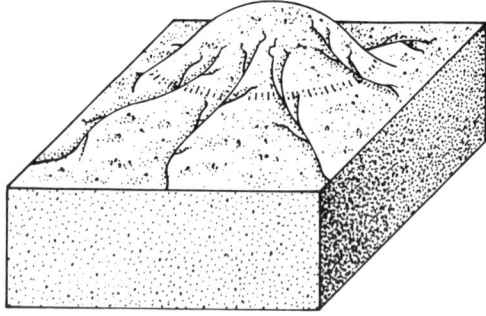

Rectangular

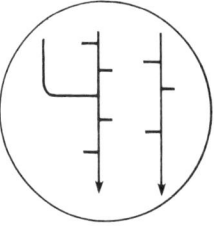

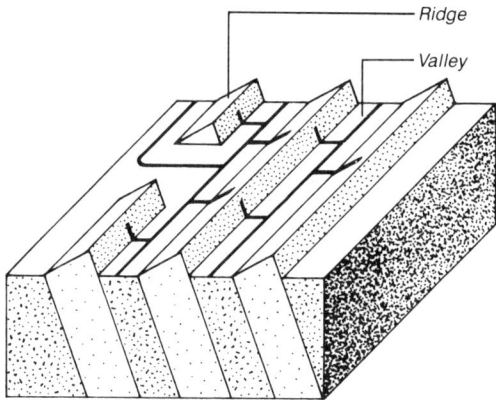

Ridge

Valley

Trellis

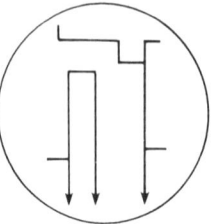

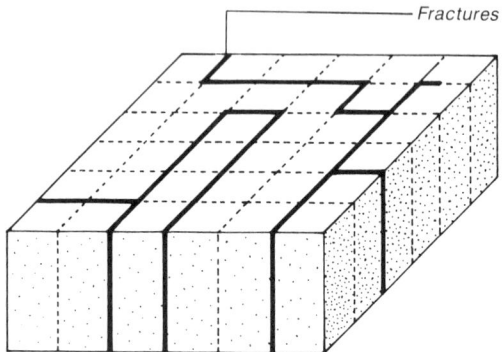

Fractures

tributaries parallel to the main stream; it resembles the stems of a vine on a trellis

d. In a *rectangular drainage pattern,* the tributary streams join others at right angles; this pattern develops on regularly fractured rock

e. In a *deranged drainage pattern,* the streams form disordered patterns as they flow into and out of lakes or swamps

8. The development of the drainage pattern is dependent on the type of bedrock, amount of runoff, climate, and the amount of vegetation

D. Stages in landform development

1. The principle of uniformitarianism states that the processes at work on the earth today have been present in the past (weathering, erosion, and so forth) and that, given enough time, these processes could account for all present-day landforms

2. Geologists have differing opinions on how a landscape develops as a region is eroded

3. In the late 19th century, W. M. Davis of Harvard University advanced a theory (the Davis Cycle of Erosion) in which he visualized every landscape going through a series of changes from initial uplift to complete leveling of the land by weathering and erosion; he designated these changes as stages: youthful, mature, old age, and peneplain

   a. The *youthful stage* begins when a stream starts downcutting a flat land surface (walls of the canyon are almost vertical)

   b. The *mature stage* is the stage of maximum relief, when no flat interstream areas remain because of mass wasting and the rounding of the slopes by sheetwash

   c. The *old age stage* is characterized by lower and more rounded interstream areas (slopes) and side valleys

   d. In the *peneplain stage,* all irregularities are eroded away, leaving a flat, featureless plain

4. A more recent theory is that slopes undergo parallel retreat; over time, the land is worn flat to form a plane, but the slope angles remain constant

   a. The angle that the slope maintains is largely dependent on the rock type involved, the climate, and whether or not the rock strata are bent (folded), broken (faulted), or lie in flat horizontal layers

   b. Rock type dictates the angle of the slope; for example, a resistant rock layer, such as a sandstone, could form an almost vertical cliff, whereas a less-resistant shale layer would produce a more rounded slope

   c. Climate also influences the slope angle; a humid or wet climate results in smoothly rounded topography, whereas an arid climate results in sharp, angular topography because of the scarce rainfall and slow chemical weathering

   d. Rock structures, such as folds, produce ridges with unequal slopes (one steep and one gentle) and faults can produce valleys and mountain ranges with unequal slopes (see Chapter 14, Earthquakes and the Earth's Interior, and Chapter 15, Deformation, Mountain Building, and Continental Crust, for more details)

   e. Horizontal rock layers with varying resistance produce stairlike topography of cliffs and slopes

Study Activities

1. Differentiate between velocity and discharge.
2. Explain how the gradient of a stream is determined.
3. Describe how a stream widens its valley.
4. Discuss the three ways in which a stream transports its load.
5. Describe how a meander can become an oxbow lake.
6. Explain the difference between a drainage system and a drainage basin.
7. Outline the stages in the development of a stream valley.
8. Sketch the four drainage patterns and explain how each particular type forms.

10

Groundwater

Objectives

After studying this chapter, the reader should be able to:
- Define groundwater and explain its origin.
- Discuss the relationships among the water table, the zone of aeration, and the zone of saturation.
- Distinguish between porosity and permeability.
- Describe the formation of a perched water table.
- Discuss what a good reservoir rock is and why it is necessary in aquifer formation.
- Explain how an artesian well flows without being pumped.
- Describe a cone of depression, and discuss where and how it forms.
- Identify the ways in which water wells become contaminated.
- Explain the circumstances under which springs are created.
- Discuss how caves and sinkholes develop, and explain how they can create karst topography.

I. Properties and Characteristics of Groundwater

A. General information
1. *Groundwater* is water stored in the open spaces of underground rocks and unconsolidated material
2. Groundwater starts as part of the hydrologic cycle, which describes the continual recycling of water from the oceans, through the atmosphere, to the continents, and back to the oceans (see Chapter 1, Overview of Physical Geology, for further details)
3. When rain falls, it can run into streams, lakes, or oceans, evaporate, be absorbed by plants and *transpired* (the release of water vapor into the atmosphere by plants), or enter the ground and become groundwater
4. How much precipitation actually ends up as groundwater depends on such factors as climate, topography, amount of vegetation, and rock type
5. Groundwater moves downward by gravity and stops when all the soil and rock openings are filled or when it encounters an impermeable rock layer (one that will not let the water flow through)
6. Groundwater does not move in channels like surface streams, but rather moves more or less uniformly (barring impermeable rocks or confining structures) through the earth toward sea level

7. Groundwater may be our most important natural resource because it represents the largest reservoir of freshwater readily available to humans (approximately 22% of the world's water supply)
   a. After glaciers, groundwater is the most abundant source of freshwater on earth
   b. Groundwater is the chief contributor to stream flow because it sustains streams during periods when rain does not fall

B. Factors influencing groundwater availability

1. Climate is the chief factor controlling the amount of water available to replenish groundwater supplies
   a. Water does not fall uniformly over the earth's surface; the amount of rainfall received by an area can range from hundreds of inches per year (as in rain forests) to virtually no rainfall over the course of several years (as in some desert regions)
   b. Runoff is rapid in arid regions, where soil layers are poorly developed and vegetation is scant, with few or no plant roots to hold the meager soil; here, water runs off before it can filter into the earth and become groundwater
   c. In arctic climates, groundwater may be trapped as permafrost or permanently frozen ground
   d. In temperate climates, thick soil profiles limit runoff; here, until the soil becomes saturated, the rainwater seeps into the ground and becomes available for groundwater
2. Topography and rock type also influence the amount of water available as groundwater
   a. In mountainous areas, where runoff is rapid and only a thin layer of soil overlies the bedrock, little water is retained as groundwater
   b. A steady rainfall on moderate slopes composed of more easily penetrated material results in a larger percentage of water entering the ground than in the scenario described in part 1. point d. above
   c. In flat areas where soil horizons have developed, precipitation does not run off, and rainwater soaks into the ground and becomes available for groundwater

C. Subsurface zones

1. The *zone of aeration,* or the unsaturated zone, is characterized by pore spaces that are filled with air and water
2. As groundwater *percolates* (moves downward by gravity) through the zone of aeration, it reaches a zone where all open spaces in the sediment and rock are completely filled with water; this is called the *zone of saturation* (see *Subsurface zones*)
3. The upper limit of the zone of saturation is the **water table**
   a. The water table normally follows surface topography (higher on hills and lower in valleys); this replication of surface topography occurs because percolation stops about 5 km below the surface (pores are closed by compaction, increasing amounts of cement, and by the plastic flow of grains due to pressure)
   b. The water table level fluctuates during periods of high and low precipitation

Subsurface zones

The subsurface is divided into vertical zones with regard to groundwater percolation, as shown in the illustration below. In the zone of aeration, the pore spaces are partially filled with air and water. Directly below, in the zone of saturation, pore spaces are completely filled with water. The upper limit of this zone is the water table.

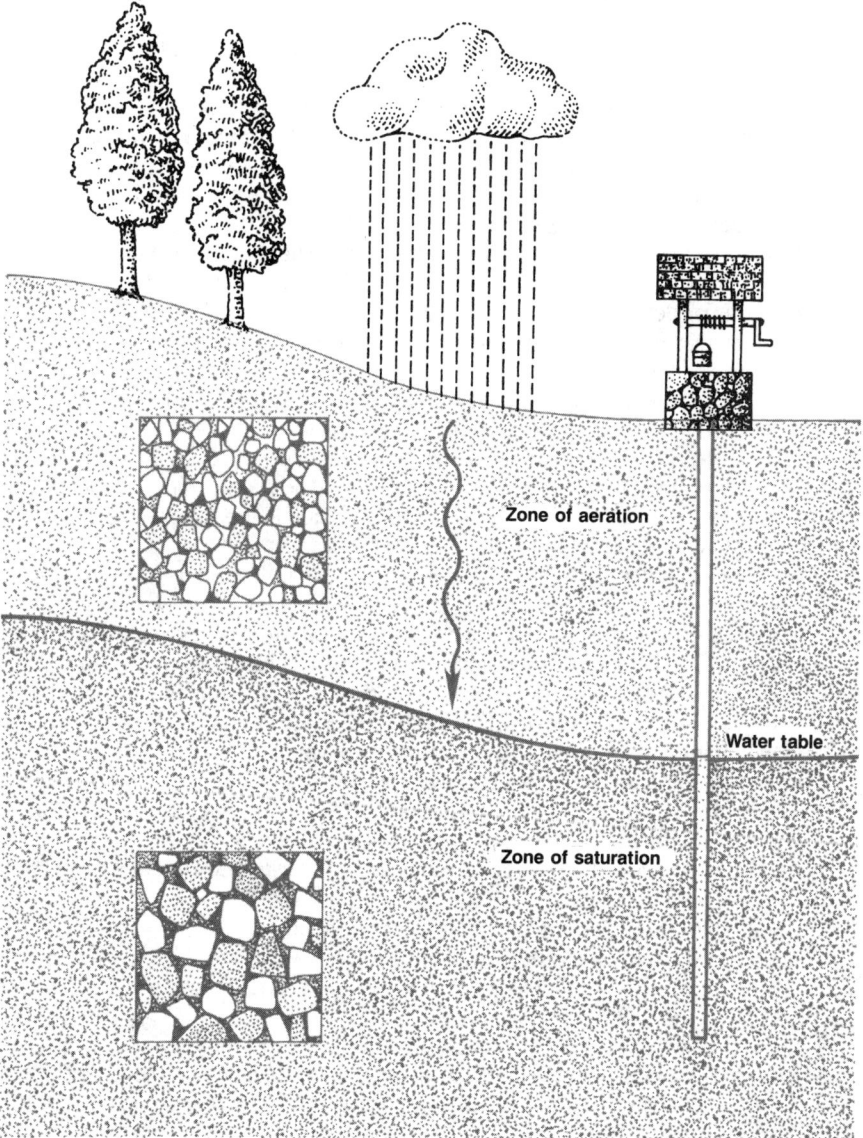

      c. A **perched water table** is a local water table that is higher than the main water table; it typically exists because of the presence of an underlying impermeable layer, such as a shale

D. Groundwater movement

1. The amount of groundwater stored in the subsurface depends on the **porosity** (the volume of pore spaces, cracks, or crevices available in the rock for holding fluid)
2. The **permeability** of a rock depends on how many of these pore spaces are sufficiently interconnected to allow fluids to flow freely through them; groundwater generally moves slowly, normally only a few centimeters per day
3. An *aquifer* is a permeable layer of sediment or rock through which water can easily move (a well must be drilled into the aquifer before the water can be extracted); well-rounded, well-sorted sands and gravels make the best aquifers because they are both porous and permeable
4. An *aquiclude* is an impermeable bed that can prevent groundwater movement

E. Groundwater outlets

1. Groundwater can reach the surface through artificial outlets, such as wells, which are able to tap aquifers
   a. A *well* is a hole drilled or dug into the zone of saturation through which groundwater can be pumped to the surface
   b. If a well pumps water too quickly from an aquifer with a slow *recharge rate* (the rate at which water from the surrounding area can replace lost water), a *cone of depression* will form (a depression in the water table immediately surrounding the base of the well)
   c. A well must be sufficiently deep so that it is not affected by a drop in the water table during drought; otherwise, it will cease to flow during a drought
2. In an *artesian well,* water rises above the level where it was initially encountered (the water may or may not reach the surface) and flows without being pumped; artesian wells form when groundwater under pressure forces the water in the well to rise above the zone of saturation, such as when water in a permeable layer is trapped between two impermeable layers
3. Natural surface outlets for ground water include various types of springs; a *spring* is a place where the water table intersects the earth's surface and a natural flow of groundwater results (see *Formation of Springs*)
   a. Springs occur when water moves along crystalline rock fractures that intersect the surface
   b. Springs occur at entrances to limestone caverns
   c. Springs occur where permeable and impermeable rock layers meet at the surface or where faults involving permeable and impermeable rocks emerge at the earth's surface
4. Swamps develop where the water table is at the earth's surface
5. Streams and lakes typically occupy areas where the land surface is below the water table; during periods of drought, the water table may drop so much that the stream or lakes become dry

F. Groundwater pollution

1. Common sources of groundwater pollution include sewage (from septic tanks and broken sewer systems) and barnyard wastes

Formation of Springs

A spring is a natural flow of groundwater emerging where the water table intersects the earth's surface. The drawings below illustrate the conditions under which springs can form. Diagram A shows how springs form where fractures intersect the surface. Diagram B shows how springs form at the mouths of caves. Diagram C shows how springs form at the contact between a permeable rock layer and an impermeable rock layer. Diagram D shows how springs form along a fault, when an impermeable layer is faulted next to a permeable layer.

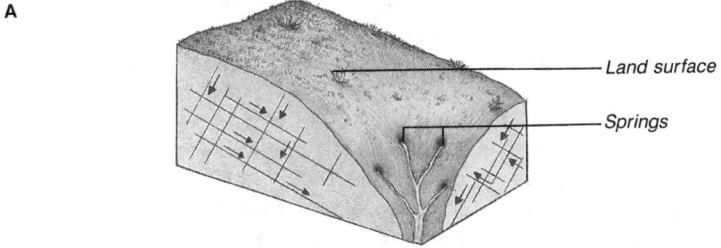

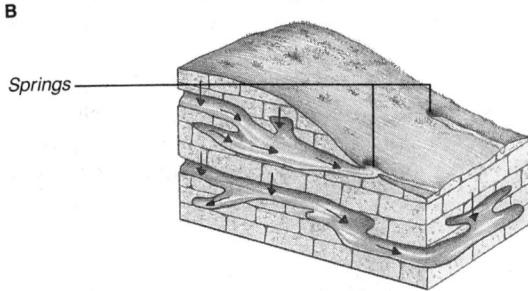

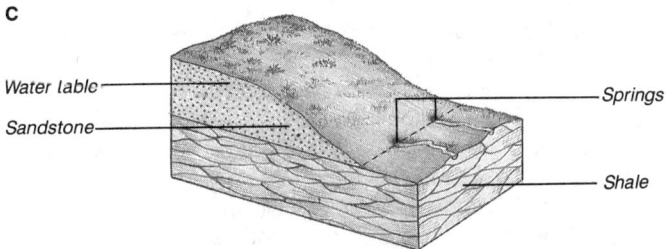

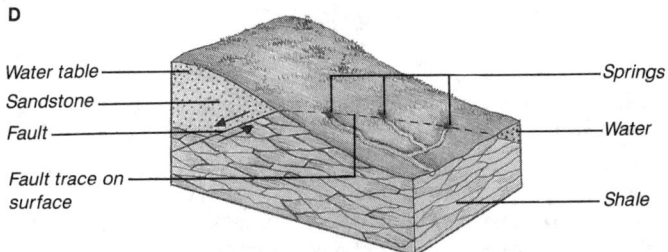

A

Land surface

Springs

B

Springs

C

Water table

Sandstone

Springs

Shale

D

Water table

Sandstone

Fault

Fault trace on surface

Springs

Water

Shale

2. Pesticides, herbicides, and fertilizers carried into the subsurface by rain or irrigation water also can pollute groundwater
3. Concentrations of household cleaning products, poisons, and paints at a dump site eventually can be dissolved by rainwater; if they reach the zone of saturation, they will contaminate the groundwater supply
4. Discarded engine oil that is dumped into the ground may eventually seep into the water table
5. Groundwater can be purified by natural processes; as water filters through the ground, harmful bacteria are removed by sediment, destroyed by chemical oxidation, or assimilated by other organisms
6. The distance groundwater must travel before it is purified depends on the permeability of the rock or sediment through which it passes
   a. If an aquifer is extremely permeable, such as those composed of coarse gravel or cavernous limestones, the groundwater may travel a considerable distance before it is purified
   b. If an aquifer is composed of sandstone or sand with lower permeabilities, the groundwater can be purified in a few tens of meters, provided water movement is slow enough
7. Overpumping of wells may create a depression in the water table, causing polluted water or saltwater (normally a safe distance from the well) to migrate to a well
8. High-level radioactive waste must be buried in sites that will provide safe storage for at least 10,000 years; the site must be located where rainfall is low (allowing for little water to percolate through the radioactive material to the water table) and where geologic activity is negligible (not in areas of faulting or volcanic eruption)

II. Effects of Circulating Groundwater

A. General information
1. Circulating groundwater dissolves rock (especially limestone and marble) along joints and fractures and creates caves and caverns
2. *Sinkholes* are surface depressions caused by either cavern roof collapse or dissolution of limestone; if numerous, sinkholes result in a hummocky surface terrain known as *karst topography*, which is characterized by a lack of surface streams and an irregular land surface
3. Groundwater can precipitate the dissolved calcium or silica it carries, forming stalactites, stalagmites, other spelothems, and various other concretions and geodes
4. Groundwater can replace material, such as wood, with silica or calcite, preserving it as solid rock known as petrified wood

B. Caves and caverns
1. Atmospheric carbon dioxide and water combine with the calcite in limestone, dissolving it and releasing calcium and bicarbonate ions into the groundwater
2. Caves and caverns form when this slightly acidic groundwater dissolves limestone along joints or bedding planes, creating a chamber
3. Limestone caverns most likely form below the water table by circulating groundwater and become exposed when the water table falls

4. Groundwater dripping through a cave roof may evaporate, thereby precipitating calcite in the form of stalactites, stalagmites, and columns, as well as many other cave deposits

    a. **Stalactites** are icicle-shaped structures that form when water dripping from the cavern ceiling precipitates calcite (calcium carbonate)

    b. **Stalagmites** are conical deposits of mineral matter that are developed by the action of dripping water; they grow upward from the cavern floor

    c. *Columns* are formed when stalactites and stalagmites grow together

C. Geothermal water

1. *Geothermal water* results when groundwater is heated by molten magma or by water circulating at great depths

2. Hot groundwater that rises to the surface may create hot springs and geysers

    a. *Hot springs* form when less dense, hot groundwater rises and flows out at the surface

    b. A **geyser** is a type of hot spring formed when the conduit through which the hot water passes is constricted and side chambers develop; this limits convection (free movement of hot water upward with colder water moving downward to take its place) and a geyser may erupt as in the following sequence

        (1) Water fills a constricted chamber

        (2) The water slowly warms and begins ascending, but bubbles of water vapor act as a stopper in the constricted part or side chamber of the geyser, resulting in superheated water (temperature rises above the boiling point without boiling because of pressure)

        (3) The upward pressure of the bubbles pushes out some of the water above it, lowering the pressure in the lower chamber

        (4) The drop in pressure causes the superheated water in the bottom of the chamber to *flash* to steam (change suddenly or violently to vapor)

        (5) The expanding vapor blasts out of the geyser in an eruption of steam and hot water

        (6) The chamber begins filling and, depending on the size of the chamber and the availability of groundwater, repeats the cycle

3. **Geothermal energy** is generated by tapping this natural steam and hot water source

D. Products of deposition

1. Deposits created by the precipitation of calcite and silica as geothermal waters emerge at the surface include *travertine* (a porous form of calcite) and *sinter*

2. *Petrified wood* is a material that forms when silica, transported by groundwater, fills in or replaces the porous organic matter in buried wood

3. *Concretions* generally are hard, rounded masses formed when calcite or silica is deposited around an organic nuclei (such as a bone or leaf)

4. *Geodes,* which are partially hollow rock bodies that have an outer shell of silica (chalcedony) and well-formed quartz or calcite crystals growing inward toward their center, also are products of groundwater deposition

Study Activities

1. Identify the factors that influence groundwater availability.
2. Draw the subsurface zones and label each part.
3. Discuss how groundwater moves.
4. Explain the difference between porosity and permeability.
5. Describe the formation of a cone of depression.
6. Illustrate three ways a spring can develop.
7. List several ways in which groundwater may become polluted.
8. Briefly explain how sinkholes, caverns, and karst topography develop.

11

Glaciers and Glaciation

Objectives

After studying this chapter, the reader should be able to:
- Explain how a glacier forms and why it moves.
- Identify localities where glaciers are found today.
- Distinguish between an alpine or valley glacier and a continental glacier.
- Describe the ways glaciers erode, transport, and deposit material.
- Explain how the landforms of alpine and continental glaciers differ.
- Explain how scientists know that glaciers have advanced and retreated in the past.
- Discuss the theories of what causes glaciation.
- Explain why the Pleistocene Epoch is known as the Ice Age.

I. Glaciers

A. General information
1. A **glacier** is a mass of moving ice that forms on land in areas where more snow falls in winter months than melts during summer months
2. Falling snow is light and loosely packed, but as the snow accumulates it begins to compact; this compaction primarily results from the weight of the additional snow layers forcing out air, and results from the thawing and refreezing of the snowflakes
3. As the weight of the accumulating snow increases, the pressure on the underlying snow causes it to recrystallize into ice; when this mass of ice becomes so large that it begins to flow under its own weight, a glacier is created
4. Snow that has been converted by compaction to granular ice is called *firn;* this granular ice undergoes further compaction and recrystallizes into glacial ice
5. Although we think of ice as a brittle substance, below about 30 m ice behaves like a viscous fluid and begins to flow like tar
6. Glaciers cover about 10% of the earth's surface and contain about 80% of the earth's freshwater
7. Glaciers are classified as either alpine (valley) or continental, based on their origin and size
   a. Alpine glaciers exist in the mountains of the western United States (as well as Alaska), in western Canada, in the Andes of South America, the Alps of Europe, in the Himalayas of Asia, and in other high mountains of the world
   b. The largest existing continental glacier is the ice sheet that almost completely covers Antarctica; the glacier covering Greenland is called an *ice*

cap (smaller than an ice sheet) rather than a continental glacier because Greenland is not a continent

B. Glacial budget and movement

1. The upper part of the glacier that is continually covered with snow is called the *zone of accumulation;* the lower part of the glacier where ice is lost (or wasted) by melting, evaporation, and calving is called the *zone of wastage*
   a. If more snow accumulates in the winter than melts during the summer, the glacier advances
   b. If more snow is lost by wastage than gained by accumulation, the glacier retreats
   c. If the amount of accumulation equals the amount of wastage during a year, the glacier is said to have a balanced budget
2. Glaciers flow down previously formed stream valleys; like the water in a stream, the greatest rate of ice movement is in the center because the valley walls and floor slow the glacier's movement via friction
3. Glacial movement is always downward (to the bottom of the valley) even if the glacier is retreating; the rate of movement is controlled by the thickness and temperature of the ice as well as the gradient of the surface over which it passes
4. Although the steepness of the slope affects velocity, an increase in temperature probably has the greatest influence on the rate of movement because higher temperatures produce meltwater, which acts as a lubricant and aids in basal sliding
5. The ice in the base of the glacier flows plastically because of high pressure; this area is referred to as the *zone of flowage*
6. The ice near the surface of the glacier remains brittle and develops cracks or crevasses as it moves; this upper part of the glacier is called the *zone of fracture* or *rigid zone*
7. A *crevasse* is an open crack or fracture in a glacier's surface
   a. Most crevasses form as a result of the greater movement of ice in the center of the glacier
   b. However, some crevasses form when the glacier passes over a cliff or an irregular bedrock surface

C. Types of glaciers

1. *Alpine,* or *valley, glaciers* are rivers of ice that descend from high mountain peaks; these glaciers are confined to mountain valleys and may belong to an interconnecting system of mountain valleys
   a. Alpine glaciers form steep, rugged landforms and U-shaped valleys with cirques at their head; a *cirque* is a bowl-shaped feature scoured at the source by the forming glacier and created largely by plucking action as ice repeatedly freezes to the bedrock
   b. Alpine glaciers originate when more rapidly moving ice breaks away from the snow or ice field; this site is commonly marked by a gaping crack known as a *bergschrund*
   c. They frequently have tributary glaciers that flow into them
   d. Alpine glaciers follow valleys originally carved by rivers, straightening and enlarging them; they do not create valleys of their own

e. A *piedmont glacier* forms when two or more valley glaciers emerge from their valleys, spread out, and form a broad sheet on the lowlands at the base of a mountain

2. *Continental glaciers* are great sheets of ice that cover large land areas (at least 50,000 km^2) and differ from valley glaciers in that they are not confined by topography; the ice in a continental glacier moves downward and outward from a central high point toward the edges of the glacier

   a. Continental glaciers are not confined to valleys; thus, they spread out in all directions with little regard for topography, decreasing the relief of the land by filling low areas with glacial drift and eroding high areas

   b. Continental glaciers are not fed by tributaries as are alpine glaciers

   c. An *ice cap* differs from an ice sheet only in its smaller size; the Greenland ice cap is an average of 3,000 m thick and covers 1.7 million km^2, whereas the continental *ice sheet* covering Antarctica covers more than 13.9 million km^2 and has a maximum thickness of approximately 4,300 m

D. Glacial erosion and transport

1. Glaciers can erode by *plucking* material from valley floors and walls

   a. Friction causes some melting at the base of the glacier

   b. The water enters fractures in the bedrock and, upon refreezing, the bedrock adheres to the glacier and is separated from the floor or walls and carried along as the glacier moves

2. Glaciers also erode by *abrasion*

   a. Rock fragments frozen to the bottom of the glacier or pushed along beneath it grind and polish all irregularities off the surfaces over which they pass

   b. The thicker the glacial ice, the more pressure on the rocks and the more effective the grinding and polishing

   c. *Rock flour* is produced by the grinding of rock into fine (silt- and clay-sized) fragments; it generally consists of particles of unaltered minerals

3. Erosional landforms associated with alpine glaciation include U-shaped valleys, truncated spurs, hanging valleys, cirques, tarns, horns, arêtes, and fjords (for illustrations of these landforms, see *Landforms Created by Alpine Glaciation,* page 88)

   a. *U-shaped valleys* are caused by glaciers moving through, straightening, and widening old river valleys

   b. Glaciers wear away ridges perpendicular to the valley, forming *truncated spurs,* which are triangular-shaped ends of ridges that have been eroded by an advancing valley glacier

   c. *Hanging valleys* are sites of former tributary glaciers that once joined the main valley glacier; when the glaciers melted, the valleys remained hanging high above the main valley floor (the tributary glaciers did not erode as deeply as the main glacier)

   d. *Cirques* are deep-sided hollows carved into a mountain at the head of a glacial valley; *tarns* are lakes that occupy cirques after the glaciers melt

   e. *Horns* are formed when a series of cirques encircle a mountain peak

   f. *Arêtes* are sharp ridges that separate adjacent glacial valleys

   g. A *fjord* is a glacially scoured valley flooded by the sea

4. Erosional landforms associated with continental glaciers include large-scale features (such as rounded topography or beaches and terraces deposited by

Landforms Created by Alpine Glaciation

Glaciers can erode landscapes by plucking material from valley walls and sides and by abrading the surfaces over which they travel. This schematic diagram shows some of the landforms that can form as a result of alpine glaciation.

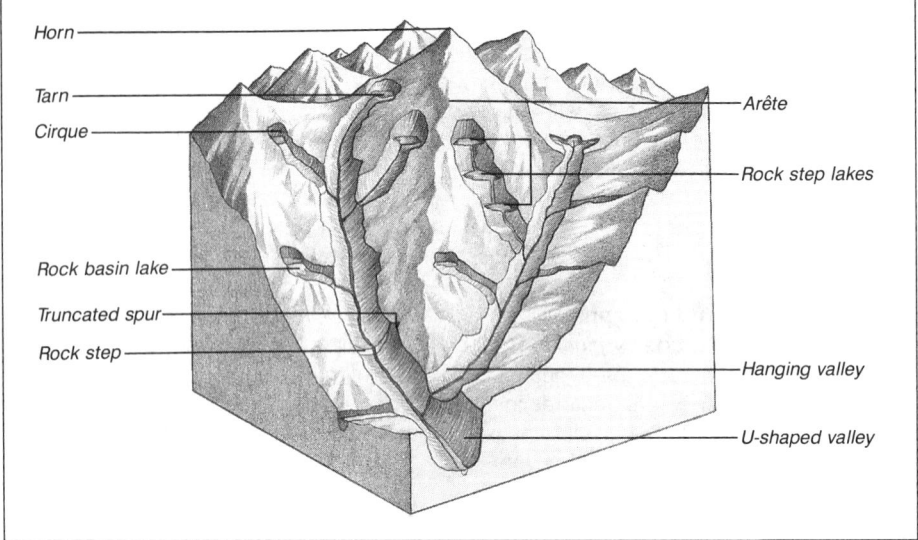

temporary glacial lakes) and small-scale features (such as polished, grooved, or striated rock surfaces)

a. Continental glaciers smooth out the topography or the landforms over which they pass, eroding mountain tops, valleys, and regions between valleys

b. Lakes that formed along the margins of continental ice sheets resulted from glacial ice damming meltwater; these temporary glacial lakes (for example, Utah's Great Salt Lake is a glacial lake remnant) left shoreline features (such as beaches and terraces) as well as fine sedimentary lake bed deposits

c. *Erratics* are large glacial boulders composed of a rock type different from any found in the area and could only have traveled the hundreds of miles from their sources on or in glacial ice

d. *Glacial striations* (scratches) and *grooves* (deep scratches and gouge marks) on bedrock surfaces are produced by the grinding action of a passing glacier and its load of sediments (the rock surfaces also may be highly polished)

5. Glaciers can transport any size of sediment that is supplied to them

6. The sediment carried by alpine glaciers is abraded and plucked from valley walls and floors; a significant amount of sediment is derived from material falling or sliding onto the glacier surface by mass-wasting processes

7. Alpine glaciers transport sediment in all parts of the ice, but the majority of it is carried in the base and sides

8. Continental glaciers or ice sheets and ice caps derive the majority of their sediment load from the land surfaces over which they move and transport it in the lower part of the glacier

Landforms Left by a Retreating Continental Glacier

Unlike alpine glaciers, continental glaciers are not bound by topography; instead of following valleys, ice sheets extend outward in all directions, leveling high areas and depositing glacial drift in low areas. This diagram illustrates some of the landforms created by the erosion and deposition of a continental glacier.

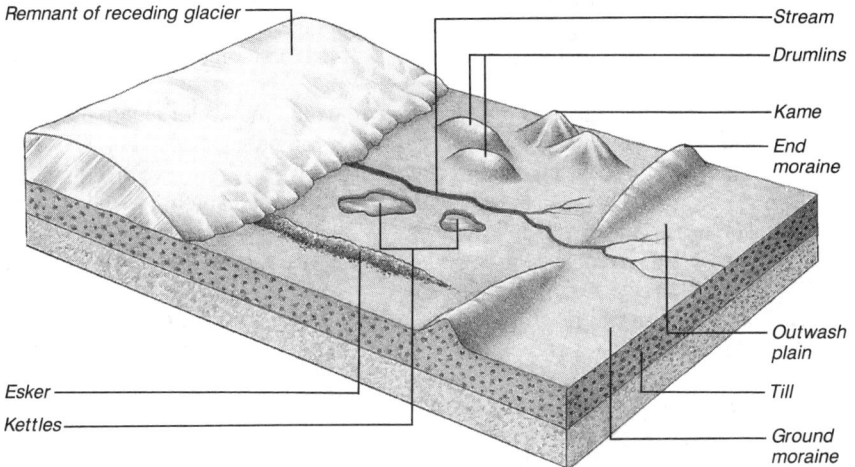

E. Glacial deposits

1. *Glacial drift* is material transported and deposited by a glacier (see *Landforms Left by a Retreating Continental Glacier*)
2. *Till* is glacial drift that is not stratified (arranged in strata)
3. *Moraines* are formed by the rock waste that is carried by the ice and deposited either at the edges of the glacier or at the furthest reaches of the glacier; these moraines are classified according to their location
   a. A *ground moraine* is the material deposited by the glacial ice as the glacier retreats; it consists of fine clay, sand, striated and rounded pebbles, and boulders
   b. A *terminal moraine,* the outermost moraine, marks the farthest advance of the glacier; the debris generally is deposited in a crescent-shaped ridge
   c. A *lateral moraine* forms from the debris accumulated on the sides of a valley glacier; the debris consists of material plucked from the valley sides or other debris that fell or slid onto the glacier's surface from the valley walls
   d. A *medial moraine* forms where two tributaries of an alpine glacier meet; it results from the union of two lateral moraines
   e. A *recessional moraine* is a ridgelike accumulation of drift deposited by a glacier along its outer margin; it is located away from the area of maximum advance of the glacier
4. An *esker* is a winding ridge of irregularly stratified sand and gravel, deposited by streams that flowed in tunnels beneath the glacial ice

5. A *kame* is a conical-shaped hill of sand and gravel that originates when sediment is deposited by heavily laden glacial streams flowing into holes in the ice near the terminus of the glacier
6. A *kettle* is a small kettle-shaped depression that forms when large masses of ice are left behind (generally buried by *outwash,* or material deposited by meltwater) during recession of ice sheets
7. A *drumlin* is an oval-shaped hill of unstratified glacial drift that probably forms when a glacier overrides and reshapes a deposit of till left by an earlier glacial advance; drumlins are shaped like inverted spoons and their gently sloping end points in the direction the former glacial ice traveled (sometimes called whalebacks)
8. *Proglacial lakes* are lakes formed from accumulated meltwater along the glacier margins; other lakes form in depressions scoured out by glaciers or where a stream's drainage was blocked by a glacier
9. Deposition in glacial lakes produces thin layers of sediment that form in alternating light and dark layers, called varves
   a. A *varve* is a glacial lake deposit that represents one year's deposition in the lake
   b. The light layer of the varve consists of silt- to clay-sized particles that form during the spring or summer months; the dark layer consists of clay-sized particles and organic matter that settle out of suspension as the lake freezes over during the winter months

II. Glaciation

A. General information
1. Evidence that glaciation occurred several times in the geologic past exists
2. In 1817, Swiss naturalist Louis Agassiz published a discourse that supported the theory that extensive continental glaciation occurred in the past in Europe
3. Agassiz and others noted that characteristic erosional and depositional features of present-day glaciers in the Alps could be found in northern Europe and the British Isles, which were far from the farthest advance of the Alpine glaciers
4. Agassiz came to North America and consulted with American geologists who had found evidence of widespread glaciation on the North American continent
5. It was proposed that the climate was colder in the geologic past, and much more of the land's surface was covered by glaciers
6. For another ice age to commence, it would be necessary for average temperatures to be lower than present-day temperatures; it is estimated that a mean annual decrease in temperature of 7.2° to 9° F (4° to 5° C) could bring on a new ice age
7. To bring on another ice age, the climate must also be sufficiently humid enough and precipitation abundant enough to ensure the accumulation of enough snow to form glaciers
8. If all existing glaciers melted, the sea level would rise between 30 and 45 m (100 and 150 feet), thereby flooding coastal cities

B. Pleistocene Epoch glaciation
1. During the Pleistocene Epoch (which began 1.6 million years ago), ice sheets and alpine glaciers were far more extensive than they are today; the Ice Age, as the

Pleistocene Epoch commonly is called, was marked by fluctuating climatic conditions that led to alternating glacial and interglacial periods
2. Four separate glacial and interglacial stages (periods when climates were as warm or warmer than today) are recognized for North America
3. Each of these glacial stages is named for the state where the deposits of that ice sheet are well exposed or were first studied, including (from oldest to youngest) the Nebraskan, Kansan, Illinoisan, and the Wisconsinan glacial stages; however, recent studies of deep-sea cores indicate that there have been at least 20 warm-cold cycles, so this four-part subdivision may need modification
4. The Ice Age began between two and three million years ago, and during that period, approximately 30% of the land's surface was covered with ice; the last glacial episode peaked about 18,000 years ago, and the ice retreated about 10,000 years ago
5. The glaciers of the Pleistocene Epoch occupied much of Canada and the northeastern and north central parts of the United States
6. Past periods of glaciation caused a worldwide drop in sea level; it has been estimated that sea level was as much as 130 m (425 feet) lower than today; land bridges developed between Alaska and Siberia, France and Britain (joined where the English Channel is today), and Indonesia and Southeast Asia
7. During the Pleistocene, numerous large lakes formed as a result of greater precipitation and overall cooler temperatures; these *pluvial lakes* formed far from the glaciers and occupied many of the basins in the western United States
8. The weight of the great ice sheets during the Pleistocene depressed the land areas, and the crust responded by sinking deeper into the mantle; however, since the glaciers retreated, the land has been rebounding to its former elevation

C. Causes of glaciation
1. A number of theories have been proposed to explain the cause of continental glaciation, including reduced solar heating, changes in atmospheric and oceanic circulation, and changes in the position of the land masses
2. The most widely accepted theory, and one that explains the cyclic nature of the glacial and interglacial periods of the Pleistocene Epoch, was proposed in 1920 by Milutin Milankovitch, a Serbian astronomer
3. Milankovitch proposed that minor irregularities in the earth's rotation and orbit are sufficient to alter the amount of incoming solar radiation, and that these irregularities occurred at intervals in the past that closely paralleled the alternating periods of glaciation and warming during the Pleistocene Epoch
4. Long-term climatic changes are thought to be a consequence of land mass changes related to plate tectonics (for example, continents moved over polar regions)
5. Short-term climatic changes that could cause warming or cooling of the earth may be related to several factors
   a. A temperature decrease could be caused by a variation in solar energy, resulting from changes in the sun itself
   b. An increase in carbon dioxide in the atmosphere (perhaps resulting from reduced vegetation) may cause a *greenhouse effect* (carbon dioxide, other gases, and water vapor allow sunlight to penetrate the atmosphere but trap heat reflected from the earth's surface, resulting in an overall increase in temperature)

c. Volcanic eruptions can block radiation from the sun, thereby cooling the earth's surface
6. Changes in oceanic circulation (brought about by shifting land masses) could have caused cooling in previously temperate areas

Study Activities

1. Describe alpine and continental glaciers, and explain how each type forms.
2. Compare and contrast the movement, landforms, and deposits of alpine and continental glaciers.
3. List four present-day theories of the probable causes of glaciation.
4. Define the following glacial landforms: cirque, horn, moraine, esker, and drumlin.
5. Identify the ways a glacier erodes.
6. Describe an area where a continental glacier occurs today.
7. List the four glacial advances recognized in North America in order of their occurrence.

12

Wind and Deserts

Objectives:

After studying this chapter, the reader should be able to:
- Define and explain the cause of wind.
- Identify the processes by which wind erodes.
- Discuss how wind transports sediment.
- Name the types of wind deposits.
- Define desert, and explain why deserts form.
- Discuss the distribution of deserts around the earth.
- Identify the characteristics of deserts.
- Describe common landforms found in deserts.

I. Wind

A. General information

1. *Wind* is the horizontal movement of air; the underlying cause of wind is the unequal heating and cooling of the earth's surface (warm air is less dense than cold air, so as warm as rises, cold air moves in to take its place)
2. Although running water is the most important agent of erosion in arid regions, wind also plays a significant role; this is because sediment in arid regions is dry and loose and vegetation scarce, and because wind is not confined to canyons so its effect is widespread
3. Wind transports fine sediment in suspension and sand-sized or larger particles in its bed load
4. Wind erodes by *abrasion* (the grinding and scraping of rock or mineral surfaces by friction and by the impact of other particles) and by **deflation** (the lifting and removal of silt- and clay-sized particles by wind)
5. Wind is more effective than water at sorting sediment by size because fine sediment can be picked up and transported for miles by wind, whereas sand-sized and larger particles are left behind
6. Windblown dust and silt-sized particles form unstratified, sheetlike deposits called *loess,* whereas sand-sized particles accumulate and form drifts or *dunes* (hills or ridges of wind-deposited sand)

B. Wind erosion

1. Wind erodes by abrasion, or sandblasting; because sand cannot be lifted more than approximately 1 m above the land surface, this abrasive effect only modifies existing landforms by etching, pitting, smoothing, or polishing them
   a. *Ventifacts* form when cobbles and pebbles, partially exposed on the desert floor, are subjected to sandblasting, which produces a flat wind-polished surface; if the rock is repeatedly turned over (probably due to the undermining of the pebble by deflation of supporting sand), another flat surface will develop
   b. *Yardangs* are elongate, streamlined ridges parallel to the wind direction; they probably form by differential erosion in which depressions, shaped like the hull of a ship, are carved out of the rock by wind
2. Wind erosion by deflation removes fine sediment
   a. A *blowout* or depression hollow can form if an area is predominantly composed of fine sediments; the resulting depressions can range in size from a few meters wide and a meter deep to several kilometers in diameter and tens of meters deep
   b. *Desert pavement* is a thin layer of closely packed, gravel-sized particles (2 mm or more in diameter); it forms when these larger fragments are left behind after the wind has removed the finer sediment (stops further deflation)

C. Methods of sediment transport by wind

1. Wind moves sediment in two ways: by suspending it in air or by rolling it along the ground as part of its bed load
   a. Suspended sediment consists of clay- and silt-sized particles, which can be carried great distances and high into the stratosphere
   b. The bed load consists of sand-sized or larger materials that are moved by **saltation,** a process by which the particles are rolled and bounced along the ground, impacting on other grains and creating a chain reaction
2. The separation (sorting) of sediment by size is quite definitive, much more so than with running water; 90% of sand grains are never raised higher than about 1 m above the ground, whereas dust-sized particles can be lifted thousands of meters
3. If the wind is strong enough, dust and sand storms occur; dust storms consist of airborne particles less than 1/16 mm in diameter and sand storms consist of moving sediment 1/16 to 2 mm in diameter

D. Types of wind deposits

1. Clay- and silt-sized material is deposited in sheetlike layers called loess; loess is thought to derive from three possible sources: glacial outwash deposits, desert, and river flood plain deposits
2. The clay- and silt-sized sediment can be composed of grains of quartz, feldspar, micas, or calcite
3. Loess-derived deposits result in fertile soils; the major grain-producing regions of the world correspond to areas of loess accumulation (for example, the Great Plains of the United States)
4. Sand-sized particles form several distinct types of dune structures that include barchan, parabolic, transverse, and longitudinal dunes (for illustrations, see *Types of Sand Dunes*)

Types of Sand Dunes

The chart below compares the features and characteristics of the most common types of sand dunes.

TYPE	IILLUSTRATION	CHARACTERISTICS
Barchan		• Crescent-shaped, with horns pointing downwind • Found in areas of constant wind and limited sand supply • Achieves height of 1 to 30 m
Transverse		• Has asymmetrical ridges transverse or perpendicular to strongest wind direction • Found in areas with strong wind and abundant sand supply • Can occur when two barchans merge
Parabolic		• Crescent-shaped, with horns pointing upwind • May have U or V shape • Trailing arms may be stabilized by vegetation • Found along coastal regions
Longitudinal		• Long, symmetrical, ridge-shaped dunes; also called seifs • Found in areas with abundant sand and a strong, constant wind

a. *Barchan,* the most common, are crescent-shaped dunes, with the horns of the crescent pointing in the same direction that the wind is blowing; they form in areas where wind direction is constant and the sand supply limited

b. *Transverse* dunes can occur when two barchans merge; they form asymmetrical ridges (transverse or perpendicular to the wind direction) in areas where winds are strong and where the sand supply is abundant

c. *Parabolic* dunes also are crescent-shaped, but the horns of their crescent (anchored by vegetation) point in the direction from which the wind is blowing; they form in coastal areas where the winds are moderate, the sand supply is abundant, and some vegetation is present

    d. *Longitudinal,* or linear, dunes form elongated symmetrical ridges of sand that are aligned parallel to the wind direction; they form in areas with strong and constant winds where the sand supply is abundant

5. Wind forms *cross-bedding,* which results from wind moving sand up a dune face or gentle slope until the grains reach the top and slide down the slipface or steep side of the dune (cross-bedding also can occur in sand ridges deposited by ocean currents on the sea floor, in sediment bars and dunes deposited by rivers in their channel, and in deltas at the mouths of rivers)
6. Wind causes the dune to move forward grain by grain, sorting the grains and sometimes causing ripple marks to form on the windward slope; moving dunes are a problem in that they may destroy agricultural land and expand desert regions

II. Deserts

A. General information

1. A *desert* is an area that receives less than 10″ (25 cm) of rain a year; although we often associate deserts with intense heat, deserts are found in cold climates as well
2. Deserts cover approximately 30% of the earth's surface and are distributed around the world in regions near the latitudinal lines 30° north and south of the equator as a result of atmospheric circulation, in areas situated in the rain shadow of mountains, within the interior regions of continents, and along tropical coasts next to cold ocean currents
3. Deserts are characterized by internal drainage, high evaporation rates, poorly developed soils, and scant vegetation
4. Desert landforms include playa lakes, playas, salt flats, alluvial fans, bajadas, pediments, inselbergs, plateaus, mesas, and buttes
5. The main agent of erosion in desert regions is rainfall; the rainfall typically is seasonal, heavy, and of short duration
6. Due to the lack of vegetation, rain spreads out in a thin layer called *sheetwash* and begins moving downhill by gravity, removing any loose surface debris along its path
7. The splash, or impact, of raindrops on the desert floor also can dislodge sediment that is later removed by sheetwash

B. Distribution of deserts

1. Some of the largest deserts of the world are products of the global circulation of air
   a. The dry regions that occur between 20° and 30° latitudes (both north and south of the equator) result from high-pressure belts of descending dry air
   b. The air under these belts becomes so warm and dry that rain seldom falls and evaporation is high (for example, the Sahara desert)
2. The *rain shadow* effect of a mountain range also is an area where deserts develop; rain shadows form when warm, moist air rising on the windward side of mountains expands, cools, condenses, and falls as rain, and when the air descends on the leeward side, it warms and becomes drier, thus creating an area of little rainfall (for example, Death Valley, in California, is in the rain shadow of the Sierra Nevada mountains)

3. Because moisture comes primarily from the oceans, areas in the interior of large continental masses have low rainfall, resulting in the formation of deserts (for example, the Gobi desert of China)
4. Deserts also can develop along tropical coasts next to cold ocean currents; as the cold air above these currents moves over the land, it warms, resulting in high evaporation and little rain (for example, the western coast of South America and the western coast of Africa)

C. Characteristics of deserts

1. Deserts receive little rainfall and typically have internal drainage (that is, they lack through-flowing streams); the internal drainage usually flows into basins (depressed areas into which streams drain)
2. Deserts are subject to heavy rainfall of short duration, which results in rapid runoff because of scant vegetation, thereby producing sheetwash and flash floods
3. Desert streams generally are heavily laden with sediment (derived from mechanical weathering), which results in rapid downcutting, producing narrow canyons with steep, vertical walls, and flat, gravel-strewn floors
4. *Desert varnish,* a shiny chemical coating of manganese oxide that frequently stains the surface of canyon walls in desert regions, is another characteristic feature (for example, Canyon de Chelly in Arizona)
5. Limestones, which are eroded easily in humid climates, stand as cliffs and ridges in desert regions
6. Mechanical weathering produces coarse and angular rock debris that is moved downhill by gravity, forming talus; deserts generally lack the rounded look of more humid climates because little chemical weathering occurs

D. Desert landforms

1. *Playa lakes,* flat areas on the floor of an undrained desert basin, appear during wet periods; evaporation of these lakes creates *playas,* dry lake beds characterized by mudcracks and chemically precipitated rocks, such as gypsum and rock salt
2. *Salt flats,* or salt pans, are created as the playa lake evaporates during dry periods; salts, formerly dissolved and concentrated in the lake water, remain as precipitates
3. ***Alluvial fans,*** somewhat similar to deltas, form where streams flow out of mountains onto basin or valley floors; in arid regions, the water seeps into the ground and sediment is deposited in a fan shape
4. *Bajadas* form when adjacent alluvial fans coalesce
5. *Pediments,* erosional surfaces cut into the solid rock at the foot of a retreating mountain, occur between mountain fronts and basin bottoms and commonly form extensive bedrock surfaces over which erosion products from the retreating mountain fronts are transported to the basins
6. An *inselberg* is an isolated, steep-sided erosional remnant of a residual hill or mountain that rises abruptly above the surrounding desert plains (for example, Ayers Rock in Australia)
7. In the Colorado plateau region of the southwestern United States, dissection of the horizontal layers of resistant rock (sandstone) that overlies less resistant rock (shale) has produced spectacular landforms, including plateaus, mesas, and buttes

a. A *plateau* is a broad, flat-topped area that has been elevated and typically is bounded by cliffs (at least on one side); the elevation is sufficient to allow deep valley cutting

b. As the plateau is dissected by streams, a *mesa,* a broad, flat-topped hill bounded by cliffs and capped by resistant rock, can form; as erosion continues to shape the mesa, it can form a *butte,* a flat-topped pinnacle of resistant rock with very steep sides

c. As erosion weakens the shale layer below the sandstone cap, the top layer breaks and tumbles to the bottom of the cliff, forming *talus* (the accumulation of broken rock at the base of a cliff), thereby decreasing the size of the mesa or butte

Study Activities

1. List two ways in which the wind erodes.
2. Illustrate the four common types of sand dunes.
3. Explain how plateaus, mesas, and buttes form.
4. Name three locations where you would expect to find deserts.
5. Explain why sheetwash is a much more effective agent of erosion in deserts than wind.
6. List four characteristics of deserts.
7. Explain why not all deserts are hot.

13

Ocean Coastlines and the Ocean Floor

Objectives

After studying this chapter, the reader should be able to:
- Discuss the forces that generate waves.
- Describe how a water particle moves within a wave.
- Identify the way waves erode, transport, and deposit material along the shore.
- Describe how depositional features are formed.
- Explain how coastal features are formed by wave erosion.
- Describe what midocean ridge systems are, where they occur, and how they form.
- Explain what fracture zones are and how they relate to the midocean ridge systems.
- Describe the stages in the formation of a coral atoll.

I. Ocean Waves and Beaches

A. General information
1. Most waves derive their energy and motion from wind that blows over the water, causing the water's surface to oscillate
2. Waves also can be generated when earthquakes or volcanic eruptions cause displacement of the sea floor; these are called *seismic sea waves* or **tsunamis**
3. In waves, it is the wave form that moves forward, not the water itself; individual water particles within a wave move in a nearly circular path during the wave's passage and return to almost their original position after the wave passes
4. This oscillatory or circular motion diminishes at a depth equal to about one-half the wave length
5. A *beach* is the sloping shore of a body of water that is washed by waves and tide; it typically is covered by sand or pebbles
6. Beaches are sometimes called *rivers of sand* because so much sand can be moved parallel to the shore for long distances
7. The process that moves the sand (as well as coarser material) along the shore is called *longshore drift* or *beach drift*
   a. Waves move up the beach at an angle, carrying the sand
   b. As the waves withdraw back toward the water, they move at right angles to the shoreline, causing the sand to migrate down the beach

B. Characteristics of waves
1. The top of a wave is called a crest; the area between crests — or the lowest point between wave crests — is called a *trough*

2. *Wave length* refers to the horizontal distance between two successive wave crests
3. *Wave height* refers to the vertical distance from the highest point of the wave (the crest) to the lowest point of the wave (the trough)
4. As the wind speed increases, the height and steepness of waves also increase
5. The bottom of the wave, or *wave base,* is located at a depth equal to one-half the wave length; it is the point at which orbital motion of the water particles ceases — that is, no wave motion is felt below this depth (for an illustration of how a water particle moves within a wave, see *Waves and Water Movement*)
6. *Wave period* refers to the time between the passage of two successive wave crests at a fixed point
7. The three factors affecting wave height, length, and period are wind speed, the length of time the wind has blown, and the distance the wave has traveled across open water
8. As a wave moves shoreward and enters shallow water, the base of the wave is slowed by friction with the sea floor; however, the wave crest, which is still advancing at the original speed, becomes too steeply sloped and plunges forward as a *breaker*
9. The energy contained within a wave is expended by breaking on the shore
   a. The collective turbulent action of the breaking waves is called *surf*
   b. The area in which turbulence takes place is the *surf zone,* or zone of breakers
10. When waves leave a stormy area and continue on, they gain in length but decrease in height, and become *swells;* waves near shore can be a mixture of swells (waves generated by a storm some distance away) and waves created by local winds

C. Wave-coastline interaction
1. Waves seldom approach the shore head on; most approach at an angle
2. As a wave approaches the shore, the part of the wave nearest the shore touches bottom and slows first, whereas the part of the wave that is still in deep water continues forward at its regular speed
3. *Wave refraction* (bending) is the process whereby the portion of a wave in shallow water slows, causing the wave to bend and align itself with the shore
4. Wave refraction strongly influences where and to what degree erosion, sediment transport, and deposition will take place
5. *Longshore currents* are currents that flow parallel to the shore and develop because the waves approach the shore at a slight angle despite refraction
6. *Rip currents* are narrow currents that flow straight out to sea through the surf zone, returning to the sea the water pushed ashore by breaking waves
   a. Rip currents form when waves strike roughly parallel to the shore, pile up water onto the shore, and return to the sea
   b. Rip currents that are in a fixed location along a shore generally are located over channels or hollows on the sea floor
   c. Where waves strike the shore at an angle, water along the shoreline builds up until the excess moves out to sea through the surf zone; rip currents of this nature can migrate in the direction of the longshore current
   d. Rip currents can be located by looking for areas where the water is discolored by sediment being picked up in the surf zone and carried seaward, by the appearance of surface ripples on a very calm day or a different pattern

Waves and Water Movement

Wave action may be illusory in that it is the wave form, rather than the water, that moves forward. Individual water molecules within a wave move in a circular path as a wave passes and resume their approximate original position afterward. This illustration shows the basic movement of a wave and the movement of water particles within the wave. Note that water movement slows considerably at a depth equal to about one-half the length of a wave.

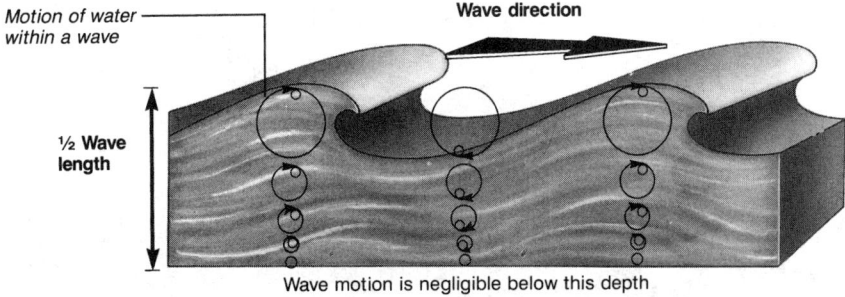

Motion of water within a wave

Wave direction

½ **Wave length**

Wave motion is negligible below this depth

of surface ripples, or by a foam line formed where early breaking waves meet the outgoing rip current
  e. Swimmers caught in a rip current can escape by swimming parallel to the shore until they are out of the current, rather than swimming against the current, which can easily tire them

D. Wave erosion, transport, and deposition
  1. Waves erode by *impact, abrasion,* and *hydraulic action*
    a. Waves sometimes strike the shore with great force, which can easily erode loose sediment
    b. Material carried in the wave (such as sand and gravel) can abrade and grind coastline rocks
    c. Waves erode by hydraulic action, which involves the pressure exerted when air and water are forced into cracks and crevices in the rocks, prying them apart
  2. Waves transport material onshore, offshore, or along the shore
    a. Waves can pick up sediment from the sea floor and move it to the shore
    b. Waves can move material offshore and deposit it in deeper, less-turbulent water
    c. Waves transport material along the shore via *longshore drift* or *beach drift*
  3. Waves deposit material in different ways (see *Shoreline Features*, page 102)
    a. They deposit material onshore to create beaches
    b. They can deposit material offshore that is still attached to the shore, as baymouth bars, spits, and tombolos
      (1) *Baymouth bars* form when longshore currents and longshore drift deposit a ridge of sediment across a bay, cutting it off from the ocean
      (2) *Spits* are fingerlike ridges of sand that project from the land into the mouth of an adjacent bay
      (3) *Tombolos* are ridges of sediment that connect an island (former island) to the shore or to another island
    c. Waves also can deposit material offshore that is not attached to the shore, as barrier islands and wave-built terraces

Shoreline Features

Waves move material onshore, offshore, or along the shore; the structures created vary in their location, attachment to the mainland, and permanency. This schematic illustration shows some of the shoreline features that occur as a result of wave erosion, transport, and deposition of materials.

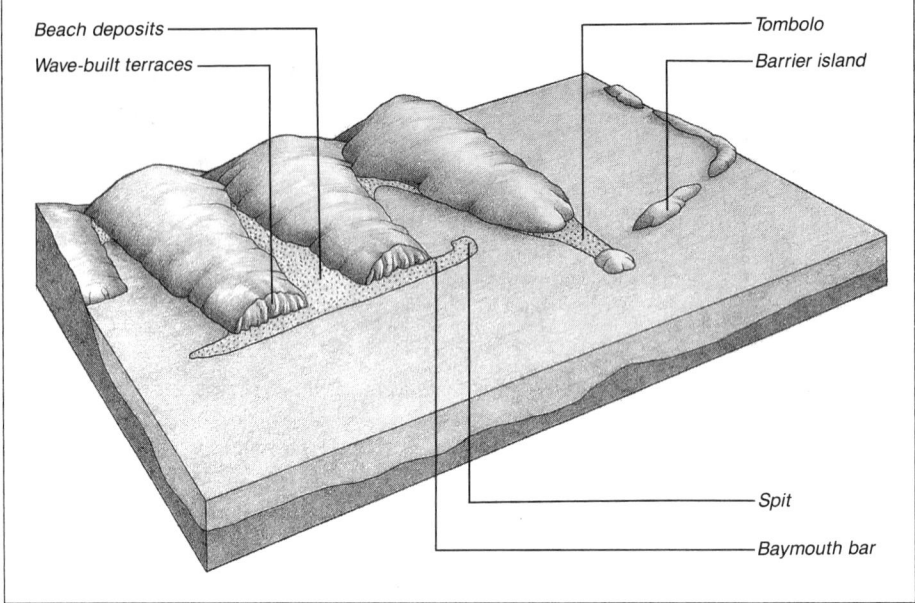

(1) *Barrier islands* are emerged ridges of sand that form parallel to the shore; for example, the Atlantic coast and the Gulf of Mexico coast have extensive barrier islands that range in size from 1 to 5 km wide and 15 to 30 km long)
   (a) Barrier islands can be former spits separated from the mainland by wave erosion
   (b) Barrier islands may be created when turbulent waters or storm waves pile up sand that has been scoured from the bottom
   (c) Barrier islands may be former sand dune ridges that originated along the shore during the last glacial period when sea level was lower; as sea level rose, the area behind the dune complex became flooded
(2) *Wave-built terraces* (composed of sediment) result from the deposition of material derived from eroded headlands

E. Characteristics of beaches

1. A *beach* is a strip of sediment (composed of sand-sized or larger particles) extending from the low-water line to the line of vegetation
   a. The sediment transported to the sea by rivers is the chief source of beach sediment
   b. Other sources of beach sediment include sediment derived from the erosion of bedrock along the coast and from the erosion of sea cliffs

   2. Beaches composed of sand- or gravel-sized sediment probably are the best
      known; however, pebble- to boulder-sized sediment can exist along shorelines
      of high-energy waves (such as areas of rapidly eroding sea cliffs)
   3. The beach surface is divided into foreshore and backshore regions
      a. The *foreshore* of a beach generally is the area covered during high tide; the
         beach face is the steepest part of the foreshore and is the area exposed to
         wave action
      b. The *backshore* is the area above the normal high-tide mark; it generally is
         dry, except during storms
      c. The backshore is composed of a *berm,* a wave-deposited sediment platform
         that slopes landward
   4. Beach sand is subject to constant motion, which is affected by the seasons
      a. Beach sand is removed during winter months when high, short storm waves
         erode material from the beach and deposit it offshore (for example, Disap-
         pearing Sands beach on the island of Hawaii)
      b. During summer months, this beach sand is redeposited when long, low
         waves wash sand from deeper water onto the beach
   5. *Offshore bars* form as sand is removed from the beaches and settles in less-
      turbulent water

II. Coastlines

A. General information
   1. The *coastline,* or shoreline, is the line of intersection between the sea and the
      land
   2. Coastlines can be areas of submergence or emergence, depending on whether
      sea level rises or falls relative to the shoreline
   3. During the last episode of glaciation, the weight of the glaciers isostatically
      depressed the land and ocean waters moved inland; however, as the glaciers
      melted, the land rebounded to its preglacial elevation and the marine waters re-
      treated; *isostasy* is the concept that the earth's crust is *floating,* in gravitational
      balance, upon the asthenosphere (see Chapter 15, Deformation, Mountain
      Building, and Continental Crust, for further details)
   4. Coastlines can be warped downward by tectonic forces, causing ocean water to
      move landward
   5. Coastlines can be advancing (material is being deposited, extending the land
      seaward), or retreating (coast is being worn away by the force of impacting
      waves)
   6. Artificial (man-made) barriers and structures can alter coastlines

B. Submergent and emergent coastlines
   1. **Submergent coastlines,** also called drowned coastlines, are created because
      the sea level has risen or the land has subsided; they are irregular coastlines
      characterized by estuaries, headlands, sea cliffs, marine or wave-cut terraces,
      wave-cut platforms, bays, sea caves, sea arches, and sea stacks
      a. *Estuaries* are funnel-shaped inlets of the sea that form when a rise in sea
         level flooded (drowned) the mouth of a river; the banks of the former river
         valley form new headlands and valleys form bays

b. *Headlands* are rocky promontories of land jutting out into the ocean around which wave energy is concentrated

c. *Sea cliffs* form as the rocky headland is eroded landward

d. *Marine* or *wave-cut terraces* are horizontal benches of rock formed beneath the surf zone in front of retreating (eroding) cliffs

e. A *wave-cut platform* or terrace is a horizontal bench of rock beneath the surf zone that is formed as a coast retreats (moves landward) by wave erosion

f. *Bays* form low areas or coves between headlands and are sites of wave deposition

g. *Sea caves* are cavities formed by wave action along zones of weakness in a cliff face

h. *Sea arches* are bridges of rock left above an opening eroded in a headland where wave action removed weak spots of rock

i. *Sea stacks* are the erosional remnants of more resistant headland material, left behind as islands when the headland is eroded landward or when the top of the sea arch collapses

2. Irregular coastlines are straightened as headlands are eroded landward and sediment derived from the headlands (and longshore drift) fills in bays

3. **Emergent coastlines,** created because the sea level has fallen or uplift of the land has occurred, generally are straighter than submerged coastlines and are characterized by uplifted marine terraces and wave-cut cliffs

a. Coastlines previously covered with ice can emerge, forcing the sea to withdraw from the land

b. Coastlines can become emergent if they are subjected to uplift due to tectonic forces, which also causes the sea to withdraw

C. Tides

1. *Tides* are the fluctuations of the ocean's surface with respect to the coast, in response to the gravitational pull of the moon and the sun

2. There are two high tides and two low tides daily; these tides can range from a few cm to more than 15 m (highest when the tide moves into an estuary or narrow inlet)

3. As the moon revolves around the earth (every 28 days), the earth is rotating on its axis (once every 24 hours); as the moon passes, the ocean waters on the side nearest the moon are attracted to the moon, causing the water to bulge

4. It takes 50 minutes longer each day for the moon to return to the same position as it was the day before, therefore the high tide occurs 50 minutes later each day (for example, high tide occurring at 12:00 one day would occur at 12:50 the next and so on)

5. As high tide occurs on the side nearest the moon, another bulge (or high tide), this one due to inertia, occurs on the opposite side of the earth

6. A *flood tide* occurs when the tide is rising and the nearshore area is flooded; an *ebb tide* occurs when the tide begins to recede

7. Both the sun and the moon exert a gravitational pull on the earth; however, because the moon is much closer, its influence is greater

a. When the moon and the sun are aligned, their gravitational forces operate together to produce a higher than normal tide called a *spring tide*

b. When the moon and sun are at right angles to one another, a lower than normal tide, called a *neap tide,* occurs

8. Tides generally have few modifying effects on shorelines (except during storms or hurricanes, when the effects are enhanced)

D. Human impact on coastlines
1. Humans have erected barriers to alter coastlines; these include jetties, groins, and breakwaters
   a. *Jetties* are walls built perpendicular to the shore to protect the entrances to harbors; sand may be eroded from one side of the jetty and deposited on the other
   b. *Groins* are short walls built to prevent sand from being moved along the beach
   c. *Breakwaters* are walls built parallel to the shore to absorb the force of large breaker waves and protect the entrances to harbors; however, sand eventually builds up behind the wall, filling it in and defeating the purpose of the breakwater
2. Beaches change to stay in equilibrium with the waves that strike them; when humans interfere with the constant (and natural) repositioning of the sand, this disturbs the equilibrium of the beach and the beach will erode and deposit material to regain equilibrium

III. The Ocean Floor

A. General information
1. *Oceanic crust* underlies the ocean basins; it has a basaltic composition and ranges in thickness from 5 to 10 km
2. The *midocean ridge* is the most outstanding feature on the ocean floor, stretching about 40,000 miles through all oceans
3. The midocean ridge extends above sea level at several places, including Iceland on the mid-Atlantic ridge and the Galapagos and Easter Islands on the East Pacific Rise
4. Continental-ocean basin margins (where continental and oceanic crust meet) consist of two types: a passive continental margin and an active continental margin
5. Submarine mountains (volcanic peaks) are found scattered on all ocean floors; the greatest number and density have been identified in the Pacific ocean, and the majority of them are associated with midocean ridges and the fractures that extend at right angles from them
6. In the tropics, coral atolls form in association with submarine volcanoes

B. The midocean ridge system
1. The midocean ridge system also is called a spreading center or a divergent plate boundary (the place where two oceanic plates are moving away from each other); it marks the area where upwelling magma adds new oceanic crust to these diverging plates
   a. Magma rising from the earth's mantle wells up along the midocean ridges
   b. As new crust is added, the older crust moves away symmetrically on either side

c. The top of the ridge is marked by a rift valley, or **graben,** a down-dropped block bounded by normal faults (see Chapter 15, Deformation, Mountain Building, and Continental Crust, for further details)
2. Ocean floor sediments thin near the ridges (presumably because the crust is younger there) and thicken near continental margins
3. Oceanic crust is geologically young; no rocks older than approximately 200 million years have been found, and these are in areas where older crust is being consumed in subduction zones
4. Transform faults (where plates slide past each other) cut the midocean ridges and are the result of magma being added to the length of the ridge system at different rates
5. Extensions of these faults create *fracture zones* up to 6,000 miles long, extending at right angles from the ridges; these fracture zones represent major lines of weakness in the earth's crust

C. Continent-ocean margins

1. A *passive continental margin* — one that is not associated with a tectonic plate boundary — consists of a continental shelf, continental slope, continental rise, and abyssal plain
   a. The *continental shelf,* which is underlain by granitic crust, is the flooded extension of the continent; it has a gentle slope of about 1° and a depth of 200 m and is located between the shoreline and the continental slope
      (1) The width of the shelf varies from a few km on the Pacific coast of North America to almost 500 km off Newfoundland; the depth of the shelf at its outer edge can vary from 100 to 200 m below sea level
      (2) During glacial periods, the continental shelves were alternately covered by ocean waters or were dry land
   b. The *continental slope,* found between the continental shelf and the continental rise (or an oceanic trench), has a steep slope (averaging 3° to 6°) and may be cut by numerous submarine canyons
   c. The *continental rise,* which rests upon the oceanic crust, is found between the continental slope and the abyssal plain; it is a gentle incline (about 5°) with smooth topography (which may be cut by submarine canyons) that extends down to the abyssal plain at a depth of about 5 km
   d. The *abyssal plain* is the flat region of the ocean floor formed by the deposition of turbidity currents and pelagic sediments
   e. *Submarine canyons* are V-shaped valleys that wind across the continental shelf and down the continental slope to the sea floor
      (1) Some submarine canyons are thought to be submerged river valleys formed when the continental shelves were exposed by the drop in sea level during the last glacial episode (many submarine canyons are located near present-day land canyons and rivers)
      (2) Swift-moving turbidity may carve the deeper parts of the submarine canyons
   f. **Turbidity currents** are bottom-flowing currents, laden with suspended sediment, which move swiftly down subaqueous slopes
2. *Active continental margins,* those associated with subduction zones (where oceanic crust is consumed), consist of a continental shelf and a continental slope that ends in a deep ocean trench

Development of a Coral Atoll

As seen in the illustrations below, a coral atoll develops from a fringing reef as a volcanic island submerges or the sea level rises. As the volcanic island erodes, the fringing reef grows
upward. Eventually, the volcano drops below sea level, leaving the coral atoll resting on the
flanks of the submerged island.

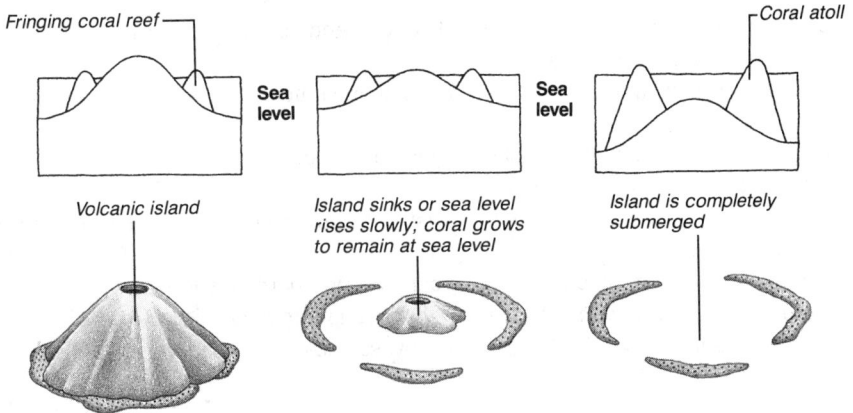

Fringing coral reef

Coral atoll

Sea level

Sea level

Volcanic island

Island sinks or sea level
rises slowly; coral grows
to remain at sea level

Island is completely
submerged

      a. A *trench* is a narrow, elongate depression found on the sea floor; according
to the theory of plate tectonics, a trench marks the site along which oceanic crust is being consumed (commonly associated with island arcs on
the ocean floor)

      b. *Island arcs,* curved chains of andesitic volcanoes rising from the deep sea
floor, typically occur between an oceanic trench and a continent (for example, the Aleutian Islands); they are thought to be formed by the partial
melting of oceanic crust as it is subducted into the mantle

D. Additional features of the ocean floor

1. Basaltic volcanoes (submarine mountains) are found singly or in lines associated
with hot spots on the ocean floor; they include seamounts and guyots (for information on how these volcanoes form, see Chapter 16, Plate Tetonics)
   a. *Seamounts* are submarine mountains that rise 1,000 m or more above the
   sea floor; they occur singly or in lines over hot spots
   b. A *guyot* is a flat-topped seamount that was once a volcanic island; its flat top
   formed when wave action eroded the top of the island away as it sank below sea level (dredging of dead coral from the tops of guyots is evidence
   of such subsidence)
2. An *atoll* is a ring of closely spaced coral islands that surround a shallow lagoon
and, in turn, are surrounded by the deep water of the open sea
3. Atolls are thought to form in stages (see *Development of a Coral Atoll*)
   a. A fringing coral reef begins growing near the shore of a volcanic island
   b. The volcano ceases eruptive activity and begins sinking (as a result of
   isostasy); as the island sinks, the coral continues growing upward to maintain its required distance from the surface (if growth is not rapid enough,
   the water will become too deep for the coral to survive)

c. Eventually, the island is completely submerged, and all that shows on the surface is the ring of coral with its base resting on the flanks of the sunken volcano

Study Activities

1. Describe the anatomy of a wave (height, length, and period, and the like)
2. Explain what causes wave refraction.
3. List three artificial structures that interfere with the natural movement of material along the coastline.
4. Explain what causes a rip current, how you could recognize one, and how you could escape if you were caught in one.
5. Compare emergent and submergent coastlines and describe at least two features of each type.
6. Describe two ways in which submarine canyons are thought to form.
7. Illustrate a passive and an active continent-ocean margin, labeling the continental shelf, continental slope, continental rise, abyssal plain, deep sea trench, and island arc.

14

Earthquakes and the Earth's Interior

Objectives

After studying this chapter, the reader should be able to:
- Discuss the nature and primary cause of earthquakes.
- List some of the geologic effects caused by earthquakes.
- Describe the three types of seismic waves.
- Explain how P waves, S waves, and the shadow zone reveal the structure of the earth's core.
- Identify the two layers in the earth that are separated by the Mohorovicic discontinuity.
- Differentiate between the focus and epicenter of an earthquake, and explain how each is located.
- Explain what a seismograph is and why the arrival times of P waves and S waves must be recorded by three separate stations to determine an earthquake's location.
- Discuss how the earth's magnetic field operates and identify its probable source.

I. Earthquakes

A. General information
1. An *earthquake* is the trembling or shaking of the earth caused by the sudden release of energy, typically resulting in displacement of rocks along faults
2. *Seismology* is the study of the vibrations produced by earthquakes
3. A *seismograph* is an instrument designed to detect, measure, and record vibrations produced by earthquakes
4. *Stress* is the force per unit of area applied to rocks (or other materials) within the earth's crust
5. The *elastic rebound theory* explains how rocks snap back or rebound after stress is released
6. *Faults* are breaks in the earth along which movement has occurred

B. Geographic distribution of earthquakes
1. Earthquakes are localized at certain places on the earth, primarily along tectonic plate boundaries; the most common sites are the circum-Pacific and the Alpine-Mediterranean-trans-Asiatic belts (see Chapter 16, Plate Tectonics, for more details)
2. Earthquakes are most prevalent in an area circling the Pacific ocean, called the circum-Pacific belt or the *ring of fire;* in this area, deep-focus earthquakes are

caused by oceanic plates being subducted under continental or other oceanic plates

3. The **Benioff zone** is the line of intermediate-focus (70 to 300 km deep) to deep-focus (more than 300 km deep) earthquakes that marks the angle of plate descent along a converging plate margin

4. Shallow-focus (less than 70 km deep) to intermediate-focus earthquakes are associated with convergent plate boundaries and mark the site where two continents collide (for example, the Alpine-Himalayan mountain belt formed when India and Asia collided)

5. Shallow-focus earthquakes are associated with transform plate boundaries, where two plates are sliding past one another (such as the San Andreas fault in California)

6. Shallow-focus earthquakes also are associated with divergent plate boundaries or spreading centers, where new crust is being added to the lithosphere by rising magma (such as the mid-Atlantic ridge)

C. Causes of earthquakes

1. Rupture of the earth along faults is the primary cause of earthquakes

    a. Stresses within the earth cause rocks to become deformed; when the strength of the rock is exceeded, the rocks break and faults occur

    b. Elastic rebound typically occurs when faults that have a slow creeping movement lock, building up stress on both sides of the fault until it breaks loose with a jerk and snaps back; the vibrations characteristic of an earthquake are the rocks elastically returning to their original shape (see *The Elastic Rebound Theory*)

2. Earthquakes also can be triggered by volcanic eruption, signaling the movement of magma within the earth

3. Man-made explosions caused by blasting or bomb detonation also can produce earthquakes

4. Geophysicists create earthquakes so that they can study information obtained by seismic waves reflecting off hidden structures; such structures may provide information about an area into which petroleum might migrate (see Chapter 17, Earth's Resources, for more details on petroleum exploration)

D. Classification of seismic waves

1. *Seismic waves* are produced when rock breaks; these waves of energy travel through the earth, causing the ground to tremble and shake during an earthquake

2. The speed at which the seismic waves travel through the earth (faster through denser material and slower through liquid or less dense material) reveals information about rock type

3. Three types of seismic waves exist: **P waves** or primary waves; **S waves** or secondary waves (P and S waves are sometimes referred to as body waves); and *surface waves* (produced when P and S waves intersect the surface)

    a. A P wave, also called a push or compressional wave, is the fastest seismic wave; the P wave arrives at the seismic station first and can travel through solids or liquids

    b. An S wave, also called a shear or secondary wave, moves material perpendicular to the direction of travel, thereby producing shear stresses in the material it moves through

The Elastic Rebound Theory

The elastic rebound theory suggests that the sudden release of energy that has been stored in elastically strained rocks results in movement along a fault, causing an earthquake. As a result, the strained rocks return to their original position. In diagram A, the stored energy, or stress, is operating on the rock. In diagram B, additional stored energy causes the rocks to become deformed. In diagram C, the release of the stored energy causes the fault to slip suddenly and the rocks to break.

A

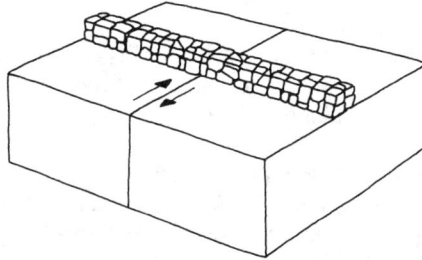

B

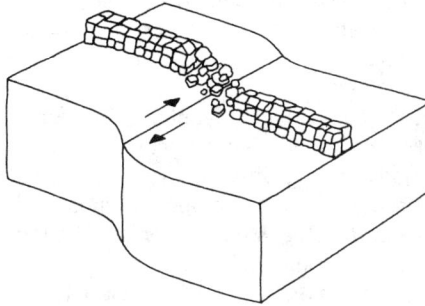

C

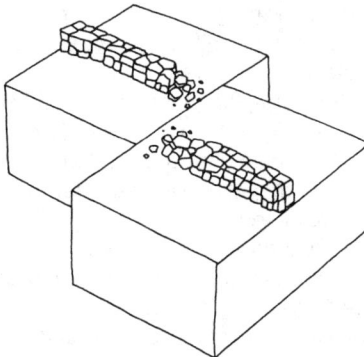

(1) Energy is lost through this side-to-side motion, causing the S wave to travel slower than a P wave
(2) Named because it is the second wave to arrive at a seismic station, an S wave can travel only through solids
c. Surface waves, also called L waves or long waves, are produced when P and S waves intersect the earth's surface; these waves are responsible for

the damage to buildings and other structures that occurs during an earthquake
4. The energy released by an earthquake radiates in all directions from its source,
the focus; the *focus* of an earthquake is that point within the earth where the
earthquake actually occurs
5. The *epicenter* is the point on the surface of the earth directly above the focus
6. Because seismic waves travel at different speeds, they arrive at the seismic
station at different times
   a. The exact arrival time of each wave can be determined by a time scale on
   the seismograph
   b. The difference between the arrival times of the P and S waves is a function
   of the distance of the seismograph from the focus (the greater the distance, the greater the difference between the arrival times of the P and S
   waves)
7. Although the distance to an earthquake's focus can be calculated from the
seismograph recording, the direction cannot be determined; all that can be as-
certained from one seismic station is that an earthquake occurred within a certain radius of the station
8. To locate the epicenter, data from three seismic stations must be obtained; a
circle is then constructed around each station (the radius of the circle is the distance to the focus from the station) and the point where the three circles over-
lap is the epicenter
9. An earthquake's intensity and magnitude are measured using two different scales:
the Mercalli scale and the Richter scale
   a. An earthquake's *intensity* is measured by the severity of the effects people
   experience and by the damage buildings and other structures sustain
   b. On the *Mercalli scale,* the intensity is ranked from Roman numeral I through
   XII; the higher numbers indicate greater degree of damage
   c. An earthquake's *magnitude,* which refers to the actual amount of energy released during the earthquake, is a measure of the maximum amplitudes
   recorded on a seismograph at a distance of 100 km; the amplitude is the
   height of the peak inscribed by the pen moving up and down on a seismograph as the seismic waves pass
   d. On the *Richter scale,* the magnitude is ranked from 0 to 8.5; the higher the
   number, the stronger the earthquake (amount of energy released)
      (1) The Richter scale uses a logarithmic scale to express magnitude
      (2) Each unit of magnitude increase represents a 30-fold increase in
         energy released
   e. People generally are more concerned with an earthquake's intensity than its
   magnitude because an earthquake that occurs far from a populated area
   may cause little damage even though it has a high magnitude; conversely,
   an earthquake of low magnitude located near a city may cause considerable damage and loss of life

E. Effects of earthquakes
1. The effects of an earthquake can cause numerous geologic changes, such as
cracks or fissures in the earth resulting from movement along faults, subsidence or uplifting of the ground, and mass movements (such as slumps, landslides, and avalanches)

2. Earthquakes under ocean basins can produce *seismic sea waves,* or **tsunamis,** which can reach heights of 50 to 100 feet (15 to 30 m) and travel at speeds of up to 450 miles (725 km) per hour; when these waves strike land, flooding and loss of life and property generally result

3. Fire is another hazard during earthquakes, particularly because broken water mains inhibit fire fighting

4. Pipes, cables, roads, and buildings can be torn apart during earthquakes

5. If subsidence of the ground occurs near the ocean, flooding and possible drowning may result

6. Where unconsolidated sediment is saturated with water, earthquakes can cause a phenomenon known as *liquefaction;* in this situation, normally solid, stable soil turns to liquid and is unable to support buildings (they sink into the ground, tilt, or fall over)

7. Methods of predicting earthquakes include studying microearthquakes, measuring the tilt of the earth and movement along known faults in earthquake-prone areas, and studying animal behavior
   a. One of the surest ways of predicting earthquake activity is to analyze the historic records of regions that have had previous earthquakes; by noting the frequency and intensity of a region's past earthquakes, reasonable estimates may be made as to when future ones may occur
   b. *Microearthquakes* are small earthquakes that typically precede the main shock; by noting an increase in these microearthquakes, it may be possible to predict when an earthquake is imminent
   c. Tiltmeters record small changes in the angle of the ground surface that are believed to result from increasing rock pressure; tiltmeters normally are placed on the opposite sides of known faults
   d. Animals and insects become more active preceding an earthquake; so, by observing increased animal and insect activity (generally under laboratory conditions), a major earthquake may be predicted

8. Building more earthquake-resistant structures can reduce the effects of earthquakes; one way is to construct the main structural supports of the building so that the building moves as one unit and will not break apart during an earthquake

II. The Earth's Interior

A. General information
1. Sir Isaac Newton (1642-1727) calculated the earth's mass to be about 5.5 g/cm^3; however, the rocks at the surface have an average specific gravity of less than 3 g/cm^3, suggesting that the earth is not homogeneous and that much of the interior must consist of materials with a density greater than the earth's average density
2. The first to provide convincing evidence of compositional layering within the earth was Andrya Mohorovicic, a 20th-century Serbian seismologist, who found that the velocity of seismic waves abruptly increases below a depth of 50 km; this boundary separates the crust from the mantle and is known as the **Mohorovicic discontinuity**
3. A few years after Mohorovicic's discovery, German seismologist Beno Gutenberg discovered that at distances of 11,000 km to 16,000 km from an earthquake

(on a nearly spherical earth, 105° to 140° of arc from the epicenter), P and S waves do not exit at the earth's surface, thus creating a *shadow zone*

   a. Gutenberg postulated that this shadow zone represented a layer in the earth composed of different material than the mantle

   b. The shadow zone is produced by refraction (bending) of P waves caused by the abrupt change in physical properties at the mantle-core boundary and represents that portion of the earth's surface where no P or S waves arrive from the earth's interior

4. It was noted that S waves (which cannot pass through liquid) completely fade away at about 105° and do not reappear (unlike P waves); geologists then concluded that at least a portion of the core is liquid

5. A 40% decrease in the velocity of P waves entering the core also points to the existence of a liquid outer layer; P waves do not increase in velocity again until they reach the discontinuity that marks the inner core boundary

6. In 1936, the last layer of the earth's interior was predicted; however, the size of the inner core was not known until underground nuclear tests in Nevada produced echoes from seismic waves, which bounced off the boundary of the inner core (P waves were found to accelerate through the inner core, indicating that this part of the core is solid)

7. Our knowledge of the earth's interior is based on observations of seismic waves and, to a lesser extent, from information obtained from studying meteorites

B. Composition of the earth

1. The earth normally is depicted as consisting of four concentric layers that differ in composition and density and that are separated from one another by distinct boundaries; these layers have been designated the crust, mantle, outer core, and inner core

2. Seismic waves are reflected off (bounced off) these layers of different density or refracted (bent) as they pass through them

3. The boundaries between these layers of different material are called *discontinuities,* and major discontinuities occur between the crust and mantle, the mantle and outer core, and the outer core and inner core; our current model of the earth's interior is based on the discontinuities between zones of different seismic wave velocities

4. Crustal material (average density of 2.7 g/cm^3 for continental crust and 3.3 g/cm^3 for oceanic crust) and mantle rock (average density of 3.3 g/cm^3) are below the average density calculated by Newton for the earth (5.5 g/cm^3); given that crust and mantle material make up about 85% of the earth's volume, the core must have a density of approximately 12 or 13 g/cm^3 in order for the average density to be 5.5 g/cm^3

5. The *crust,* or rigid outermost layer of the earth, consists of continental crust and oceanic crust

   a. The continental crust is composed of granitic rocks underlain by basaltic rocks; however, the oceanic crust, which underlies all the ocean basins, is composed entirely of basaltic rocks

   b. The crust does not have a uniform thickness, but varies from about 56 km (35 miles) under the continents to 5 km (3 miles) under the ocean basins

6. The **mantle** is about 2,900 km (1,800 miles) thick and is composed of the igneous rock peridotite, an ultramafic igneous rock which is 60% olivine, 30% pyroxene, and about 10% feldspar

a. The upper mantle, along with the earth's crust, forms the outer solid portion of the earth called the *lithosphere;* the lithosphere lies above the low-velocity zone

b. The *asthenosphere,* the part of the upper mantle below the lithosphere, is located between the depths of 100 km and 700 km; the rocks in this region are easily deformed and may be partially melted

c. The asthenosphere also is known as the *low-velocity zone* because the velocity of P and S waves decreases noticeably in this zone; the partially molten material of the asthenosphere moves and supports the rigid plates of the lithosphere

7. The *core* of the earth is divided into an outer liquid zone and an inner solid zone composed of a mixture of iron with some nickel and a little sulfur or silicon; a metallic core would have the correct density to fit model calculations needed to account for the earth's mass

   a. The *outer core* is about 2,270 km (1,407 miles) thick and is believed to be liquid; evidence to support this liquid core model include the fact that S waves cannot penetrate the outer core and P-wave velocity slows through it

      (1) The reason that the outer core is liquid while the inner core is solid is thought to be the outer core's sulfur content (estimated at about 12%); sulfur depresses the melting temperature (iron-sulfur mixtures melt at a lower temperature than pure iron)

      (2) P waves show a 40% decrease in velocity when they enter the outer core, which would be expected if the outer core was liquid

   b. The *inner core* boundary (discontinuity) occurs at a depth of approximately 5,120 km (3,179 miles) and marks the area at which P-wave velocity increases

8. The study of meteorites provides additional evidence of the composition of the earth's interior

9. Meteorites are thought to have formed at the same time as the earth, and the materials of which they are composed have the densities required to fit model calculations for the earth's mantle and solid inner core; meteorites are of two main types: stony and metallic

   a. *Stony meteorites* predominantly consist of olivine and pyroxene (peridotite is the common name for rocks of this composition), the same material that exists in the earth's mantle

   b. *Metallic meteorites* are composed primarily of iron and nickel, the same material that is thought to make up earth's solid inner core

C. Temperature and pressure in the earth's interior

1. Heat emanating from the earth's interior comes from three possible sources: the remnant heat left from when the earth cooled from a molten mass, the heat generated during the decay of radioactive minerals, and the heat rising from the molten liquid core

2. The *geothermal gradient* is the rate that the earth's temperature increases with depth (approximately 25° C/km or 75° C/mile); however, the geothermal gradient is useful only to a depth of about 100 km because, below that depth, temperature increases drop markedly and the gradient in the mantle is thought to be only 1° C/km

3. Evidence for this increase of temperature with depth is experienced in deep mines and deep drill holes
4. The temperature does not continue to increase all the way to the earth's center; it also is influenced by pressures and melting points of material
5. Higher temperatures weaken rock, whereas higher pressures strengthen them; this combination of high temperature and high pressure at lower depths causes the rocks to behave plastically
6. Rocks within the earth remain solid because their melting point (the temperature at which the rock or mineral melts) rises with increasing pressure
7. Temperatures are estimated to be 800° to 1,200° C at the base of the crust, 4,800° C at the mantle-core boundary, and 6,600° C at the outer core-inner core boundary; the temperature at the earth's center is calculated to be approximately 6,900° C
8. Upper limits for temperatures of the earth's interior are based on the melting points of rocks assumed to form each layer, taking into account the melting points of minerals under high pressure
9. Pressure increases with depth because of the weight of the overlying rocks
10. Minor discontinuities within the earth's mantle may be due to structural changes in the minerals at lower depths, caused by increasing pressure (for example, olivine, the most abundant mineral found in peridotite, is transformed into the denser mineral spinel at a depth of about 400 km and spinel is transformed into the mineral perovskite at a depth of about 700 km)
11. *Heat flow* is the escape of earth's internal heat at the surface; heat flow from continental crust and oceanic crust is approximately the same
12. *High heat flow* is detected over areas of active or recent volcanism (for example, midocean ridges and island arcs) and *low heat flow* typically is recorded over deep ocean trenches

D. The earth's magnetic field

1. A *magnetic field* is the region of influence of a magnet; magnetic fields (such as that of the earth) cause magnetic compass needles to line up in the direction of the field
2. The earth's magnetic field is much like that of a bar magnet and may be represented by lines of force that emerge from south-seeking magnetic pole and enter north-seeking magnetic pole
3. The earth's magnetic field is thought to be due to electromagnetic currents generated by convection in the fluid outer core and from the different rotation speeds of outer liquid core and mantle material
4. A magnetic compass needle points to the magnetic north pole, which is located about 11.5° from the geographic north
5. Magnetism recorded in rocks, such as basalt, show both the latitude and the direction of magnetic north at the time the rock formed; *paleomagnetism* is the study of the earth's ancient magnetic fields, and rock magnetism has proven to be one of the most important pieces of evidence for plate tectonics (see Chapter 16, Plate Tectonics, for further details)
6. The direction of the earth's magnetic pole is recorded in magnetic minerals at the time the igneous rocks in which they were formed crystallized; most basalts (rocks that make up oceanic crust) contain magnetite, a magnetic mineral that aligns with the magnetic pole at the time the basalts crystallize

7. By taking measurements of the direction and angle of the magnetic field of rocks on all the continents, the field movements relative to one another can be interpreted

8. *Magnetic reversals,* or the reverse in polarity of the earth's magnetic field, has occurred numerous times in the past; evidence of these reversals is recorded in oceanic crustal material associated with midocean ridges

9. During periods when magnetic reversal occurs, the magnetic lines of force leave the earth at the north magnetic pole and enter the earth through the south magnetic pole (normally, the lines of force leave the south magnetic pole and enter the earth through the north magnetic pole)

10. *Magnetic anomalies* are local variations in the expected normal magnetic readings obtained by a *magnetometer,* an instrument that measures the earth's magnetic field and its changes; when geologists prospect for an ore body, the vertical magnetic intensity is measured

    a. *Positive magnetic anomalies* give readings of magnetic strength stronger than those normally found in the surrounding area; for example, they can indicate a hidden iron ore body or geologic structure

    b. *Negative magnetic anomalies* give readings of magnetic strength weaker than those normally found in the surrounding area; for example, they can be produced by a down-dropped fault block, such as a graben

Study Activities

1. Briefly describe the movement associated with P waves and S waves.
2. Illustrate what happens to P and S waves as they move through the earth's interior.
3. Explain how the Mercalli scale measures the intensity of earthquakes.
4. Discuss how the Richter scale measures the magnitude of earthquakes.
5. Explain why people are more concerned with an earthquake's intensity than its magnitude.
6. Describe three methods currently used to predict earthquakes.
7. Draw a sketch of the interior layers of the earth, giving their thicknesses and relative compositions.
8. Compare the lithosphere with the asthenosphere, and describe the layers of the earth in which they are found.

15

Deformation, Mountain Building, and Continental Crust

Objectives

After studying this chapter, the reader should be able to:
• Explain the difference between stress and strain.
• Differentiate between a fault and a joint.
• Describe the forces that cause faulting and list the types of faults.
• Discuss why most folding and faulting occurs at continental margins.
• Describe how folds are formed.
• Explain the meaning of diastrophism.
• Differentiate between orogenic and epeirogenic types of movements.
• Discuss how the theory of isostasy explains why ocean basins are lower than continents.

I. Deformation

A. General information

1. Rock deformation is defined as any change produced by tectonic forces in the original form or volume of rock masses; folding, faulting, and flow are common modes of deformation
2. The amount of deformation depends on the strength of the rock formation and the length of time the forces have been at work
3. A *force* is defined as that which changes or tends to produce a change in the state of a body; the two forces producing these changes are stress and strain
   a. *Stress* is the force that acts on a rock body and tends to deform or change its size or shape
   b. *Strain* is the change in volume or shape of a rock body in response to stress
4. **Diastrophism** refers to all movements of the earth's crust resulting from vertical and horizontal changes of position and the deformation of rocks
5. The types of deformation structures include gentle tilts or warps, folds, joints or fractures, and faults

B. Stress and strain

1. The principal types of stress are compressive stress, tensile stress, and shear stress

a. *Compressive (or compressional) stress* tends to push together material on opposite sides of a real or imaginary plane; compressive strength is the load per unit of area under which a rock mass fails by shear or fracturing (it involves the pressure needed to deform or crush an object permanently)

b. *Tensile (or tensional) stress* tends to stretch or pull apart material on opposite sides of a real or imaginary plane; tensile strength is the ability of a material to resist a tensile stress

c. *Shear stress* is caused by forces parallel to, but in the opposite directions of, each other; shear strength is measured by the maximum shear stress a rock can withstand, based on an original cross-sectional area that can be sustained without failure

2. The compressive, tensile, and shear strength of an object dictates its ability to resist deformation; for example, granites have a very high compressive strength (20,000 to 30,000 lb/inch^2), but a low tensile strength (600 to 1,000 lb/inch^2)

3. An object's strength is influenced by temperature and pressure; a higher pressure tends to increase strength, whereas a higher temperature tends to decrease it

4. The principal types of strain are elastic and plastic strain

a. *Elastic strain* is nonpermanent deformation, and the body returns to its original shape as soon as the force is removed; the elastic limit is the maximum stress a material can undergo without being permanently deformed

b. *Plastic strain* deforms an object permanently and does not involve failure by rupture or fracture; plastic strain also is defined as any permanent deformation throughout which a rock maintains its essential cohesion and strength

c. At depths where temperatures and pressures are high, rocks behave plastically and, although they are still in the solid state, they begin to bend into folds

d. Rocks in the lithosphere are brittle and, instead of behaving plastically, they break or fracture

C. Deformation produced by tilting and warping

1. The earth's crust over time has been tilted, warped, uplifted, and depressed

2. Gentle tilting or warping of the crust produces domes and basins

a. A *dome* is an upwarp in which beds dip away in all directions from a central point (domes also can form as parts of large anticlines or by the intrusion of igneous rocks doming up overlying sedimentary rock layers)

b. A *basin,* is a depression, or downwarp, in which the strata dip toward the center from all sides (basins also can result from folding or erosion)

3. Warping of the earth's crust may cause gentle swells and troughs, which may be barely noticeable

D. Deformation produced by folding

1. Rocks begin to respond to stress by folding (a *fold* is a bend in layered bedrock); the type of fold produced depends on the forces at work and the amount of time the forces have been operating

2. Fold types include anticlines, synclines, overturned folds, and recumbent folds (see *Types of Folds,* page 120)

a. An **anticline** is the upwarp or arched portion of a fold

b. A **syncline** is the downwarped or trough portion of a fold

Types of Folds

Rocks respond to stress by folding; the type and duration of the stress determine the characteristics of the fold. This illustration shows the most common types of folds found in rock strata: a monocline, an anticline, a dome formed by an anticline, a syncline, an overturned fold, and a recumbent fold.

Monocline

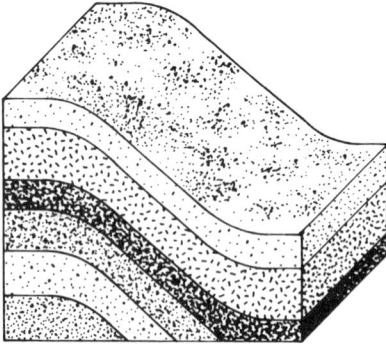

Anticline

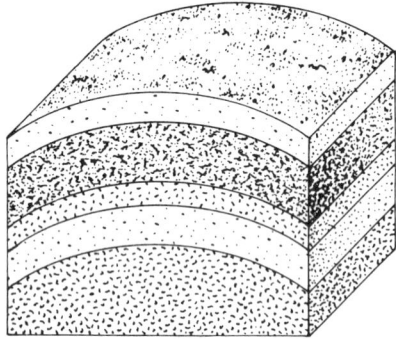

Dome formed by anticline

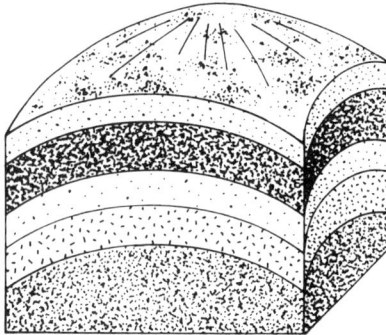

Syncline

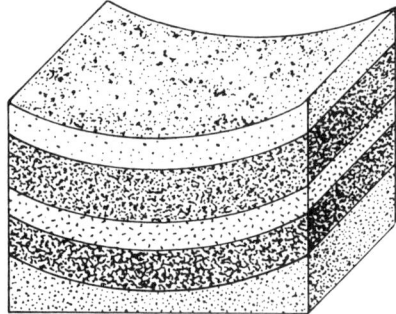

Overturned fold

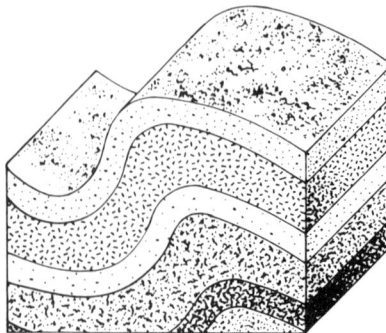

Recumbent fold

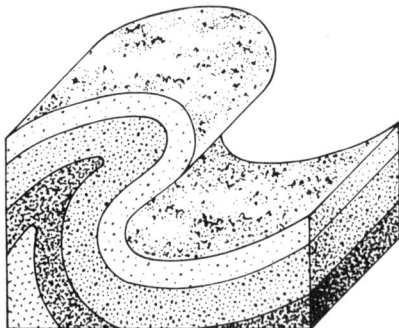

      (1) The *limb* of the fold is that portion shared by both the anticline and syncline

      (2) The *axial plane* divides the fold into equal (symmetrical) parts, whereas the *fold axis* delineates the maximum curvature of the fold

   c. An *overturned fold* is one in which the axial plane is inclined more than 90°

   d. A *recumbent fold* is one in which the axial plane is more or less horizontal

3. Folding typically occurs in the earth's interior where rocks behave plastically and fold instead of break, as they do in the lithosphere

E. Strike and dip

1. To find the extent and direction that rock strata have been moved from their original position, the relationship between the surface of an inclined bed and an imaginary horizontal plane must be determined; this is done by determining the strike and dip of a bed

2. The **strike** gives the compass direction of a line formed by the intersection of an inclined plane with a horizontal plane

3. The **dip** is the angle of inclination or tilting of the strata from the horizontal, and is measured perpendicular to the strike

4. The dip and strike together define the position of a particular layer of rock with respect to the horizontal surface and compass direction

5. Horizontal beds have no dip and therefore no strike; therefore, strike and dip are used only when dealing with tilted or folded beds

F. Joints and faults

1. If rocks cannot release stress by folding, they break or fracture forming joints

   a. *Joints* are breaks in bedrock where no movement has occurred; they generally are produced by tensile or compressive forces

   b. Joints produced by tensile forces tend to pull an object apart, typically resulting in parallel fracture patterns; those produced by compressive forces result in fractures that meet at right angles to one another

   c. Joints also can be produced by expansion and contraction

      (1) *Expansion joints* form curved surfaces on exposed outcrops of plutonic igneous rock bodies (such as batholiths) and are caused by the release of pressure when the rock mass has been uplifted and the overlying rock strata has been removed by erosion

      (2) *Columnar joints* refer to a type of jointing that breaks the rock into columns, forming a more or less clearly defined hexagonal pattern; these types of joints are caused by contraction of cooling magma and are formed in basaltic lava flows and other extrusive and intrusive rocks

   d. Rocks with right-angle fractures produced by compression (such as granite and limestone) make ideal building stone because they are easy to quarry

   e. Joints frequently occur in sets (groups of more or less parallel joints)

2. Fault movement along fractures occurs when the competence (inherent strength) of the rock has been exceeded

   a. A **fault** is a break in the earth along which movement has occurred

   b. Fault movement may be normal or reverse and the position of the fault surface may be vertical, horizontal, or inclined

     c. A fault may be classified according to whether its direction of movement is along the strike, down the dip, or at an oblique angle to both the dip and strike

     d. Most fault surfaces are inclined, with one block (rock mass) overlying the other (for illustrations, see *Types of Faults,* page 124)

        (1) The **hanging wall** is the overlying surface of the inclined fault plane (it refers to an old mining term used when miners hung their lamps from the surface overhead)

        (2) The **footwall** is the underlying surface of an inclined fault plane (it refers to the surface on which the miners walked)

     e. The type of fault is determined by the movement of the hanging wall relative to the footwall

        (1) A **normal fault** is one in which the hanging wall has moved down relative to the footwall; it typically is produced by tensile forces

        (2) A *reverse fault* is one in which the hanging wall has moved up relative to the footwall; it is produced by compressive forces

        (3) A **thrust fault** is a low-angle reverse fault

        (4) A *strike-slip fault* is one in which movement is essentially vertical along the direction of strike

        (5) A *transform fault,* a special variety of strike-slip fault, is one along which the displacement suddenly stops or changes form; it also forms a plate boundary that ideally exhibits strike-slip displacement (this trans-form movement is discussed briefly in Chapter 16, Plate Tectonics)

        (6) An *oblique-slip fault* is one in which has vertical as well as horizontal movement has occurred

G. Evidence of fault movement

    1. Evidence that movement along faults has occurred includes slickensides, striae, fault breccias, and fault gouge

     a. *Slickensides* are polished surfaces found on the adjacent walls of the fault

     b. *Striae* are grooves or scratches found on the adjacent walls of the fault

     c. *Fault breccias* are broken or shattered particles of rock that have been crushed by movement on the fault

     d. *Fault gouge* is fine powder that is produced when rocks are pulverized in the fault zone

    2. Where fault movement has been significant, a **fault scarp** (an escarpment or cliff) may be formed on the uplifted side of the fault

II. Mountain Building and Continental Crust

A. General information

    1. **Orogenic,** or mountain-building, forces commonly are accompanied by folding

    2. **Epeirogenic** movements involve the raising or lowering of the crust by warping, with little or no folding

    3. The major mountain systems of the world are a series of folded ranges that occur singly or in groups

    4. Evidence that uplift has occurred includes finding fossil seashells at high elevations and shoreline features that indicate emergent coastlines

   5. Evidence that the crust has been depressed includes the presence of structural basins and shoreline features that indicate submerged coastlines

B. Types of mountains

   1. Geologists believe that continental crust is composed of former mountain masses; these continental masses became enlarged by accretion of mountain belts to their margins
   2. Mountains occur singly or as part of extensive mountain belts, such as the American Cordillera, which extends from the tip of South America through Alaska
   3. Mountain systems came into being as a result of enormous forces (those which cause plate tectonics) that have folded, faulted, and deformed large sections of the earth's crust
   4. *Orogenesis* (*oros* is the Greek word for mountain) can be explained by the plate tectonics theory, and mountains are believed to be formed when plates converge and continents collide
   5. The types of mountains formed include mountains created by folding and faulting, by uplift or intrusion of an igneous mass, and by volcanic action
      a. **Folded mountains** are produced by the compression or shortening of the crust
      b. *Thrust-faulted mountains* form when compressive forces become strong enough to fracture or break the folds, thrusting them up over one another
      c. *Fault-block mountains* are produced when tensile forces produce a series of uplifted and down-dropped blocks
         (1) **Horsts** are uplifted blocks, bounded by normal faults, whose eroded remnants form mountains
         (2) **Grabens** are down-dropped blocks, bounded by normal faults, which form valleys (see *Types of Faults,* page 124)
         (3) These horsts and grabens form the familiar topography seen in the southwestern United States; geologists call this area the Basin and Range province or the Great Basin
      d. *Domed mountains* can be formed by uplift of an intrusive igneous mass, or **laccolith** (a mushroom-shaped intrusive rock), which bulges up overlying sedimentary strata, causing it to dip away in all directions from the top of the dome
      e. *Volcanic mountains* are produced by successive lava flows or erupted pyroclastic material (from volcanoes) building a cone-shaped structure on the earth's surface
      f. *Submarine mountains,* or *seamounts,* are volcanoes that form on the ocean floor; if the seamount is flat-topped it is called a *guyot* (see Chapter 13, Ocean Coastlines and the Ocean Floor)

C. Theory of isostasy

   1. The theory of **isostasy** (from the Greek word meaning equal standing) suggests that different masses of the earth's crust (granitic continental and basaltic oceanic crust) stand in gravitational balance or equilibrium with each other at some depth within the earth
   2. Continents are made up of less dense granitic material, and therefore they stand higher than the more dense basaltic crustal material; both float on the asthenosphere, the plastic layer in the mantle

Types of Faults

Fault movement along fractures indicates that the inherent strength of the rock has been exceeded. The classification of faults are based on the movement of the hanging wall relative to the footwall. This diagram shows the three primary types of faults, as well as two structures produced by normal faults (a graben and a horst).

Normal fault

Graben

Horst

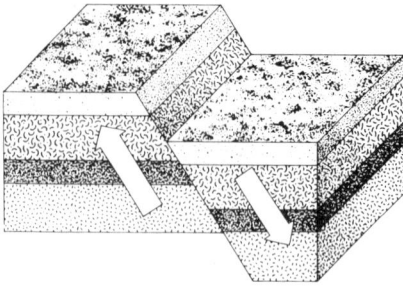

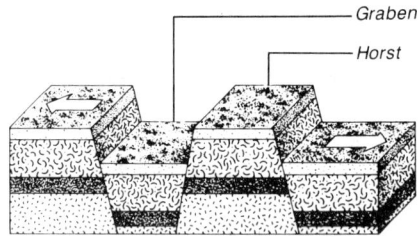

Reverse fault

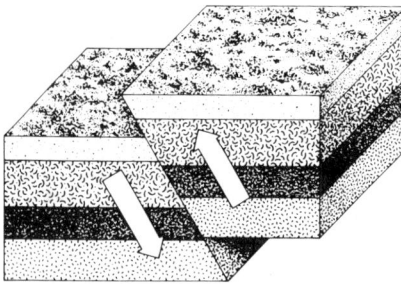

Strike-slip fault

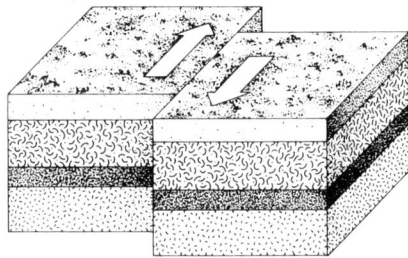

3. Mountain belts are thicker than the rest of the continental crust and so stand higher above the surface and extend farther into the dense mantle; a mountain will sink into the mantle until it displaces a volume of mantle material that equals its total weight

4. The force of gravity determines the elevation of land masses; as mountains erode away, they are continuously uplifted (to stay in gravitation equilibrium) until the deep crustal rocks (the *roots* of the mountain) are exposed

   a. The gravitational pull between an object and the earth is referred to as its weight, and gravitational attraction would be the same everywhere on the earth's surface if the earth were homogeneous; however, gravity measurements vary and these variations are called *anomalies*

      (1) A *positive gravity anomaly* indicates a force of gravity greater than the expected average; this can be caused by a buried ore body or a buried geologic structure, such an igneous intrusion

      (2) A *negative gravity anomaly* indicates the force of gravity is less than the expected average; this can be cause by a salt dome (salt is less dense than rock) or an ocean trench

    b. A gravity survey across a mountain region shows no gravity anomaly (which would be expected due to the mass of the mountain), because the mountain mass is in gravitational equilibrium

5. *Isostatic adjustment* is the concept that, if weight (such as that of a continental glacier) is added to the crust, the area will subside; conversely, when the weight is removed (the glacier melts), the area will respond by uplift

6. Evidence to support the concept of isostatic adjustment includes studies of areas that were once covered by thick continental glaciers but are now free of ice (the Hudson Bay in Canada was the site of the thickest accumulation of ice 8,000 to 10,000 years ago; since the ice melted, the area has risen approximately 330 m)

D. Continental crust

1. The earliest crust is thought to have been thin, unstable, and composed of ultramafic material; spreading centers developed along ridges where rising basaltic magma caused upwelling (the basaltic magma was generated by the partial melting of the ultramafic crust)

2. The basaltic crust, which became partially melted at subduction zones, produced volcanic island arcs of andesitic composition, and partial melting of the lower crustal andesites formed granitic magmas

3. The first granitic land masses (continents) were formed by the collision of these island arcs, which provided the nucleus or stable portion of the continents
    a. The first continental rocks are called *shields,* broad platforms of ancient rocks buried beneath younger sedimentary rocks
    b. The shields and buried portions of the shields are called *cratons*

4. The composition of this nucleus or craton (the stable part of the continents) has been changed repeatedly by igneous intrusive and metamorphic events since its formation

5. Currently, there are three hypotheses as to how continental masses became enlarged at convergent plate boundaries
    a. New material is added along continental margins as ocean basins close and continents collide (the site along which the two continents come together is called a *suture zone*); thick accumulations of sediment from both sides of the closing ocean would be compressed into folded mountain ranges that are accreted to the continent, thereby increasing their size (for example, the Himalayan mountains)
    b. At convergent plate boundaries, where oceanic crust is subducted beneath lighter continental crust, partial melting occurs and new magmas form and rise to be emplaced as volcanic and plutonic rocks along the continental margins (for example, the western coast of South America)
    c. A third hypothesis is that small fragments of crustal material collide and merge with continental margins; these fragments, called exotic terranes, are thought to be microcontinents (a *terrane* is defined as any crustal fragment whose geologic history is distinct from that of the adjoining terrane)

6. As smaller cratons collided along belts of deformation called *orogenies,* the size of the continents increased; further orogeny and accretion gave rise to the present configuration of the continental masses

7. Because new rocks are added along their margins, the older continental rocks are found in a continent's interior

Study Activities

1. Compare the compressive strength and tensile strength of granite, and explain why one would be so much stronger than the other.
2. Illustrate the major types of folds.
3. Explain why the strike and dip cannot be determined on horizontal rock layers.
4. Illustrate the three types of faults, using arrows to indicate the direction of movement along each one.
5. Name at least three types of evidence indicating that faulting of a rock layer has occurred.
6. Explain why faults are more easily detected in sedimentary rocks than igneous or metamorphic rocks.
7. List at least four ways in which mountains can be formed.
8. Briefly outline the steps in the evolution of continental masses.

16

Plate Tectonics

Objectives

After studying this chapter, the reader should be able to:
- Explain how the hypotheses of continental drift and sea-floor spreading led to the theory of plate tectonics.
- Discuss the type of magma associated with divergent and convergent plate boundaries.
- Describe the mechanism that drives the plates.
- Explain how convection cells develop.
- Discuss how hot spots develop over mantle plumes.
- Describe how active plate boundaries are located using earthquake studies.
- Explain what Benioff zones are and how they are related to subduction zones.

I. Continental Drift and Sea-Floor Spreading

A. General information
1. The theory of plate tectonics was based on the hypotheses of continental drift and sea-floor spreading
2. The theory of plate tectonics has become accepted by most geologists for two reasons: first, because an overwhelming amount of evidence supports it, and second, because it provides geologists with a means of explaining major geological phenomena, including the distribution of mountain ranges, volcanoes, ocean ridges and trenches, earthquakes, and many other processes and features
3. Alfred Wegener, a German meteorologist, generally is credited with developing the theory of continental drift
4. Evidence to support his theory included climatologic, geologic, and paleontologic data
5. The hypothesis was discredited because Wegener did not have a suitable mechanism for moving the continents around the globe
6. The theory of sea-floor spreading was proposed by Princeton geophysicist Harry Hess; he explained that the continents are not drifting, rather the sea floor is moving away from spreading centers and carrying the continents *piggyback* in the process

B. Continental drift

1. In the hypothesis of ***continental drift,*** as first proposed by Alfred Wegener in 1912, the continents were once joined in the past as a supercontinent he called *Pangaea;* the continents drifted to their present positions after Pangaea broke apart

2. The evidence Wegener used to support his hypothesis included the fit of the continents, fossil evidence, similar rock types, and similar structures found on separate continents, as well as paleoclimatic data

   a. Wegener reasoned that the continents' shapes could be fitted together like a global jigsaw puzzle

   b. He noted the presence of similar fossil plants and fossil land animals on con- tinents now separated by oceans; their similarities, he suggested, pointed to an existence on joined land masses

   c. He noted the presence of similar rock types and mountain structures, which match when the continents are reassembled

   d. Wegener also pointed out that paleoclimatic data could be used to determine if a continent had been close to the poles in the past (as evidenced by glaciation) or near the equator (as evidenced by ancient coral reefs)

3. Wegener theorized that Pangaea first separated into two large land masses

   a. The northern one, Laurasia, included North America, Greenland, Europe, and Asia (excluding India)

   b. The southern one, Gondwanaland, included Africa, South America, Antarc- tica, India, and Australia

4. Evidence that the land masses of Africa, South America, India, Australia, and Antarctica once were part of Gondwanaland includes glacial deposits, fossil plants, and fossil freshwater animals

   a. The presence of glacial deposits and striations (grooves carved into solid rock from the debris carried in the base of moving glaciers) that align when these continents are reassembled indicates that they were probably once together and covered by the same ice sheet

   b. Other evidence is the presence of the fossil fern Glossopteris, found in sedi- mentary beds overlying glacial till, on all five continents

   c. Still more evidence is the presence of nearly identical fossil freshwater ani- mals found in rocks of the same age on the southern continents (the fossil reptile Mesosaurus has been found on all five continents)

C. Sea-floor spreading

1. ***Sea-floor spreading,*** a hypothesis proposed by Harry Hess in 1960, states that new oceanic crust is being formed by rising basaltic magma at the midocean ridges; as the new crust is added, the older crust is moved away in opposite di- rections from the ridge belt until it eventually is subducted at a converging plate margin

2. Hess argued that, by this process, the continents could move with the sea floor

3. Evidence to support his theory included the relatively young age of the ocean floor, the fact that the youngest rocks are next to the ridges, and paleomagnetic studies

   a. Young-age dates from sea floor rocks (no older than 200 million years) indi- cate that the sea floor is being recycled (the oldest continental rocks found so far are 4 billion years old)

b. The youngest oceanic crust is found next to the ridges, and the oldest is found close to the subduction zones; this is another indication that new oceanic crust is added at the ridges and, over time, migrates to subduction zones where it is destroyed

c. The rocks on either side of the ridge mirror each other in age

4. Earth's present magnetic field is referred to as normal, with the north and south magnetic poles located in the vicinity of the north and south geographic poles (magnetic north is located about 1,290 miles southwest of true north); the earth's magnetic field emerges from the south magnetic pole and enters at the north magnetic pole

5. **Paleomagnetism** is the study of remnant magnetism in ancient rocks that records the direction and strength of the earth's magnetic field at the time molten rock crystallized; this information is obtained by studying the way magnetic minerals were oriented (pointing toward the magnetic north pole) at the time the igneous rocks in which they were formed crystallized

6. The ancient magnetism recorded in rocks indicates that earth's magnetic field has reversed itself numerous times in the past (for example, the earth's magnetic field would emerge at the north magnetic pole and enter the south magnetic pole, opposite of what it does now)

7. The magnetic reversals, when plotted on a map, show a pattern of parallel bands on either side of the midocean ridges; these bands match not only in polarity, but also in width

8. **Polar wandering,** or the idea that the magnetic poles have changed position throughout geologic time, arose because of the numerous directions recorded from paleomagnetic studies

a. The locations of the magnetic north pole at different times in the earth's history have been determined, and polar wandering paths have been constructed for all the continents; the lines of these paths connect the positions of the north magnetic pole for rocks of different ages

b. If the paths for each continent had been identical, this would indicate that the continents had remained stationary; however, the polar wandering paths for each continent are different, indicating that the continents have moved relative to one another

II. Plate Motion

A. General information

1. The theory of **plate tectonics** proposes that the **lithosphere,** the outer rigid portion (70 km to 125 km thick) of the earth, is divided into a number of plates; the lithosphere is composed of upper mantle and oceanic and continental crustal material

2. Below the lithosphere is the **asthenosphere,** a zone approximately 150 km thick that behaves plastically because of increased temperature and pressure; the tectonic plates slide on the asthenosphere

3. The tectonic plates, which cover the earth's surface, interact with each other along their margins; active plate margins are outlined by earthquakes

4. The rate at which the plates move is relatively slow (1 to 10 cm/year)

5. The plates are created at divergent plate boundaries, consumed at subduction zones, and carry the continents with them as they move

6. Most of the world's volcanoes occur along divergent and convergent plate margins; a belt of active volcanism around the Pacific Ocean is referred to as the *ring of fire* because of the subduction zones present now around much of this ocean

B. Plate boundaries

1. Three major types of plate boundaries exist: divergent, convergent, and transform (for illustrations, see *Types of Plate Boundaries*)
2. *Divergent plate boundaries* occur where plates are moving away from one another; this can begin in the middle of the ocean or in the middle of the continent
   a. Divergent plate boundaries also are called constructive plate boundaries because new lithosphere is created here
   b. Divergent plate boundaries can involve the interaction between two oceanic plate margins or two continental plate margins
      (1) Divergent plate boundaries involving oceanic-oceanic margins form mid-ocean ridges and are characterized by shallow earthquakes, rift valleys, and basaltic magmatism (for example, the mid-Atlantic ridge)
      (2) Divergent plate boundaries involving continental-continental margins form rift zones, which generally produce rift valleys (for example, the African rift valleys); they are characterized by basaltic magmatism, which may or may not have been contaminated by continental crustal material (a new ocean may eventually form at this widening margin)
3. *Convergent plate boundaries* occur where plates move toward one another, resulting in either the subduction of one of the plates or the collision and crumpling of two continental masses
   a. Convergent plate boundaries also are called destructive plate boundaries because old lithosphere is destroyed during subduction
   b. Convergent plate boundaries can involve the interaction between two oceanic plate margins, two continental plate margins, or an oceanic plate margin and a continental plate margin; they are associated with surface features and subsurface activity
      (1) A convergent plate boundary involving two oceanic plate margins results in subduction zones; these boundaries are associated with surface features, such as island arcs and deep-sea trenches, and are characterized by deep-focus earthquakes and andesitic volcanism (for example, islands of Indonesia and the Aleutian trench)
      (2) Convergent plate boundaries involving continental-continental margins develop when all the oceanic plate of a oceanic-continental margin has been consumed at a subduction zone, resulting in collision of two continental masses (continental crust is too light to subduct); this type is characterized by the formation of interior mountain ranges, shallow-to intermediate-focus earthquakes, and the intrusion of granitic plutons (for example, the Alpine-Himalayan mountain belt)
      (3) Convergent plate boundaries involving a continental-continental margin also form a *suture zone,* which is a line where two continental masses are joined, forming one large continental mass from two separate masses
      (4) Convergent plate boundaries involving oceanic-continental margins result in subduction zones where the oceanic crust is plunged beneath continental crust; these boundaries are associated with chains of an-

Types of Plate Boundaries

Plate boundaries are areas of interacting plate margins. Active plate margins are outlined by earthquakes, and most of the world's volcanoes are associated with divergent and convergent plate margins. This illustration shows the three major types of plate boundaries: divergent, convergent, and transform.

Divergent   Convergent   Transform

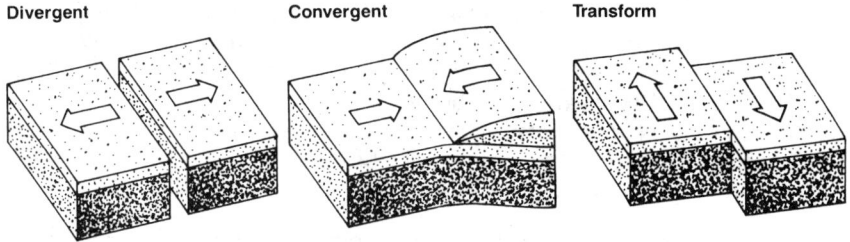

desitic volcanoes on the continents, ocean trenches, deep-focus earthquakes, and the emplacement of granitic batholiths (for example, the Cascade Range volcanoes in the northwestern United States)

(5) A common feature of island arcs and deep ocean trenches is a **Benioff zone,** a dipping seismic zone that indicates the angle of plate descent along a convergent plate boundary; the Benioff zone is marked by a line of earthquakes triggered by the descending plate

4. *Transform plate boundaries* occur where two plates slide past one another and generally end at another type of plate boundary

  a. Transform plate boundaries are formed by *transform faults,* which involve that portion of a fracture zone between two offset segments of a midocean-ridge crest

  b. Transform plate boundaries can involve two oceanic plate margins or two continental plate margins

    (1) Transform plate boundaries that involve two oceanic plate margins include faults that cut the midocean ridge system and ones that form the extensive fracture zones that traverse all basins; they are characterized by shallow-focus earthquakes and generally are not associated with any type of volcanism

    (2) Transform plate boundaries that involve two continental plate margins include faults that offset surface features; these are characterized by shallow-focus earthquakes and no volcanism (the San Andreas fault of California, probably the most famous transform fault, also is a *strike-slip fault,* involving horizontal movement in which blocks on opposite sides of the fault slide sideways past one another)

C. Cause of plate tectonics

1. The mechanisms responsible for plate motion are thought to be convection and gravity; the energy source comes from heat released by the decay of radioactive elements and probably a deeper internal heat source located within the core

2. The principal driving force that moves the plates is convection currents in the upper mantle, part of the heat transfer process from the earth's interior

3. *Convection cells* (circulating cells created by differences in temperature and density) form in the mantle when magma, which is lighter and less dense than the surrounding rock, rises at the midocean ridges, begins cooling as it moves away from the ridge, and is carried down into the mantle on the descending limb of the cell
4. The pull of gravity may move the descending plates at subduction zones (as it starts to descend, the weight of the plate itself aids in pulling it down)
5. Other mechanisms that may be responsible for plate movement are *mantle plumes;* these narrow columns of rising hot mantle rock spread out radially, break up the lithosphere, and move the plates
6. Where mantle plumes intersect the surface of the earth, *hot spots* are produced; hot spots are localized zones of melting below the lithosphere as evidenced by volcanism at the surface (for example, the Hawaiian Islands)

Study Activities

1. Outline the theories of sea-floor spreading and continental drift, listing several points of evidence for each.
2. Illustrate the three types of plate boundaries, and explain how these boundaries are located by observing surface features.
3. List and illustrate two causes of plate motion.
4. Differentiate between the lithosphere and asthenosphere.
5. Explain the type of plate boundary associated with a suture zone.
6. List the continents that were once a part of Gondwanaland, and cite evidence that they were once together.
7. Compare the concepts of polar wandering and paleomagnetism, and explain how they are related.
8. Explain why the sea floor is less than 200 million years old, while the oldest rocks found on the earth are 4 billion years old.

17

Earth's Resources

Objectives

After studying this chapter, the reader should be able to:
- Explain what is meant by a natural resource.
- Discuss the classification of natural resources and name the types of natural resources that belong to each classification.
- Describe the origin of individual types of natural resources.
- Explain how the individual types of natural resources are extracted.
- Differentiate between a renewable and a nonrenewable resource.
- Discuss the difference between a resource and a reserve.
- Explain what constitutes an ore mineral.

I. Natural Resources

A. General information
1. A *natural resource* is any material found on and in the earth that is available for human use
2. Natural resources provide all the materials we use in our daily lives, from the clothes we wear to the gasoline we put into our cars
3. The term **resource** includes all known deposits and those that may be inferred to exist but have not yet been discovered either locally or worldwide
4. A **reserve** includes all the proven resources from which the mineral or fuel can be extracted profitably with existing technology
5. Resources are either renewable or nonrenewable
   a. A *renewable natural resource* is one that can be replenished naturally, such as wood or groundwater
   b. A *nonrenewable resource* is one that once used is gone forever and cannot be replenished (for example, oil, gas, and metals)
6. The origin of these natural resources varies as does the methods of extracting them

B. Classification of natural resources
1. There are three main classifications of natural resources: energy, metallic, and nonmetallic
2. Energy resources are used for power generation; examples include coal and natural gas

3. Metallic mineral resources are those that are extracted from the crust and provide the raw materials from which many useful, if not essential, products are made (such as copper, which is used in electrical wiring, and aluminum, which is used in beverage cans)
4. The natural accumulation of most metallic elements is extremely small, so the resource must have accumulated in sufficient quantities to be extracted economically
5. Igneous processes have concentrated some of the more important accumulations of metals
6. Nonmetallic resources are those used for construction or for industrial use; they are not used as fuels or processed for metals (for example, limestone is used as building stone or in making cement)

II. Energy Resources

A. General information
1. Energy resources include the fossil fuels: petroleum, natural gas, and coal
2. Fossil fuels provide almost 90% of our energy supply at this time
3. It is estimated that the petroleum reserves in the United States will last another 25 years; after that, we will have to depend on alternate energy sources to supply our needs
4. Alternative energy sources include nuclear energy, geothermal energy, hydroelectric power, and solar and wind power

B. Types of energy resources
1. Petroleum, natural gas, and coal are known as fossil fuels because they are composed of the remains of organic materials
   a. *Petroleum* is a naturally occurring complex of liquid hydrocarbons (an organic compound) that, after distillation, yields a wide range of fuels, petrochemicals, and lubricants
   b. *Natural gas* (mostly methane), frequently associated with petroleum, is a gaseous mixture of hydrocarbons; both natural gas and petroleum originate from organic matter accumulating in marine sediment
   c. **Coal** is a readily combustible sedimentary rock formed in freshwater swamps from incompletely decayed plant material; it is rich in carbon, and brown or black in color
2. Some alternate energy sources include nuclear energy, geothermal energy, hydroelectric power, and solar and wind power
   a. *Uranium,* a metal that powers nuclear reactors, most commonly occurs as pitchblende, a black uranium oxide, or as the mineral carnotite, a complex hydrated oxide
   b. **Geothermal energy,** which provides steam to turn turbines in power plants, is derived principally from groundwater circulating near still-hot igneous masses
   c. Hydroelectric plants provide approximately 5% of our energy needs; water flowing from dammed river reservoirs turns turbines and generates electricity

    d. Solar power and wind-powered turbines also can be used to generate elec-
       tricity; however, improved methods for collecting and storing energy cre-
       ated by these energy sources is needed

C. Origin of energy resources

1. Petroleum and natural gas are concentrated underground and typically are
   obtained by drilling
2. Petroleum, as well as natural gas, requires a source rock, a reservoir rock, and a
   structural or stratigraphic trap into which the fluid or gas can migrate for extrac-
   tion
   a. A *source rock* is the geologic formation in which the oil and gas originates
   b. A *reservoir rock* is a formation, such as a sandstone, with a sufficient per-
      centage of pores (voids in which the liquid or gas can accumulate) to be of
      commercial significance and enough permeability (the ability of intercon-
      necting pore spaces to allow gases or liquids to pass through) so that the
      fluid can flow readily through it into a well
   c. A *structural trap* is the result of a structure, such as an anticline, that traps mi-
      grating oil or gas
      (1) The most common type of structural trap from which petroleum can be
          produced is the anticline; when oil and gas occur in folded sandstone
          beds, they migrate to the top of the structure and become trapped
          where the sandstone is overlain by impermeable shale layers
      (2) Faults can create oil and gas traps when the rock layer containing the
          source rock breaks and is moved adjacent to an impermeable rock
          layer
   d. A *stratigraphic trap* forms as the result of natural sedimentation rather than a
      structure; stratigraphic traps can be formed as a result of a sandstone
      lens, a pinchout, a salt dome, or a patch reef
      (1) Oil-bearing sandstone lenses trapped within an impermeable shale layer
          can be a source of petroleum
      (2) Oil-bearing sandstone layers that pinch out within an impermeable shale
          layer also can be a source of petroleum
      (3) A *salt dome* (vertical column of rock salt that has risen from a parent
          layer of salt through enclosing sediment) is impermeable and can
          form a trap if petroleum-bearing strata become upturned and faulted
          by the intrusion of the dome
      (4) A *patch reef* is a small, thick, unstratified lens of limestone or dolomite
          that is isolated from the larger reef complex by rock debris; patch
          reefs can provide a source of petroleum when oil migrates into the
          large holes in the reef, which are formed by the irregular growth of
          coral or algae
3. Coal is found where abundant plant material has accumulated and been buried in
   sufficient quantities to form an economical deposit (generally in temperate or
   tropical climates); coal occurs in beds
   a. If the coal beds are deeply buried, the deposit can be extracted through a
      mine shaft
   b. If the coal beds lie close to the surface, the overburden can be stripped off
      and the coal can be mined directly

4. The element uranium is formed by igneous processes and, because it is soluble, it is readily transportable in groundwater in the oxidized state; organic matter causes the uranium to precipitate in sandstones and shales
5. Geothermal resources are found in areas with recent igneous activity and are the result of groundwater circulating near still-hot igneous bodies in the earth's surface; wells drilled in geothermal areas can produce steam, which is used to generate electricity

D. Worldwide distribution of energy resources
1. The major areas of the United States that hold most of our petroleum and natural gas reserves include the North Slope of Alaska, as well as the oil fields of Texas, Oklahoma, and Louisiana; worldwide reserves include the Middle East, the North Sea, the former Soviet Union, Venezuela, and Mexico
2. The major areas of coal in the United States include the Appalachian fields, New Mexico, northward through the Rocky Mountains to Montana and the Great Plains of North Dakota; coal is obtained by underground and strip mining
3. Uranium in the United States is found in New Mexico and Wyoming, where it is mined by underground methods
4. The principal source of geothermal energy in the United States is the geyser fields of northern California; New Zealand and Iceland make extensive use of this energy source

III. Metallic Resources

A. General information
1. *Metallic minerals* are mined for their metallic elements
2. Numerous metals are used today; however, the most important ones are iron, copper, aluminum, lead, zinc, gold, and silver
3. These metallic minerals generally are characterized by malleability, luster, and conductivity of heat and electricity
4. Metals can occur in a variety rock types and in numerous environments
5. An *ore* is any naturally occurring material from which a mineral or minerals can be extracted profitably
6. Criteria that determine whether or not a metal can be classified as an ore include the cost of mining the metal, the prevailing market price, and the distance from the source to the market
7. Because metallic mineral deposits commonly are concentrated by igneous processes, many deposits are formed near divergent and convergent plate boundaries

B. Origin of metallic mineral deposits
1. Metallic ore deposits can be formed in several ways: by magmatic segregation as the result of crystal settling, by precipitation from hydrothermal fluids, by direct chemical precipitation from water, from the concentration of heavier minerals in placer deposits, and from concentration by weathering processes
2. *Magmatic segregation,* or *crystal settling,* occurs when minerals that crystallize in the magma move downward in the magma chamber because they are denser than the magma; sills in the Bushveldt complex of South Africa have layers of chromite (overlain by platinum) in their base

3. *Hydrothermal ore deposits,* the most important source of metallic ore deposits, originate from hot, metal-rich fluids associated with cooling magma bodies; as the magma crystallizes, metallic ions accumulate near the top of the magma chamber and the hot waters migrate into the surrounding country rock where they form contact metamorphic deposits, vein deposits, disseminated ore deposits, or hot spring deposits

   a. Contact metamorphic deposits form copper, lead, zinc, and silver deposits when fluids from an igneous intrusive mass that is rich in these metals replaces the country rock (usually limestone)

   b. Hydrothermal fluids carrying metallic ions enter fractures or cavities and joints in country rock, then cool and precipitate these metals as vein deposits; vein deposits can contain lead, zinc, silver, gold, or tin

   c. Disseminated ore deposits occur when ore minerals are distributed in extremely low concentrations in large volumes of rock either above or within a *pluton* (a body of igneous intrusive rock); metals mined from disseminated deposits include copper, lead, zinc, molybdenum, silver, and gold (called porphyry copper deposits if they are associated with a porphyritic pluton)

   d. Hot spring deposits form when hot water containing large quantities of metallic ions flows out onto the surface, cools, and precipitates metals

4. Deposits can be chemically precipitated in layers; for example, during the Proterozoic Eon, organisms (bacteria) released oxygen into seawater, causing large-scale precipitation of iron oxide and silica and resulting in the formation of banded iron deposits (a chief source of iron ore)

5. *Placer deposits* form where heavy minerals accumulate as a result of mechanical concentration by stream or ocean currents; valuable minerals found in these types of deposits include gold, tin, platinum, and diamond

6. Deposits concentrated by weathering processes include bauxite (the chief ore of aluminum), which results from the intense chemical weathering of common rock-forming minerals in tropical climates; during weathering, all the soluble elements are removed, while iron and aluminum, which are relatively insoluble, remain and become concentrated

7. Pegmatites are coarsely crystalline igneous rocks typically associated with granitic batholiths; pegmatite minerals form from the water-rich vapor phase that exists after most of the magma has crystallized and includes such elements as lithium, beryllium, cesium, and tin, along with several gem minerals, such as emerald, aquamarine, and tourmaline

C. Metallic mineral deposits and plate tectonics

1. There is a close relationship between the formation and distribution of metallic mineral deposits and plate boundaries

2. Because metallic mineral deposits are concentrated by igneous processes, it is not surprising that they are associated with plate boundaries, where igneous activity is concentrated

3. Metallic mineral deposits can be found near divergent and convergent plate boundaries; deposits also can be found in continental interiors that are sites of former convergent plate boundaries

   a. Deposits that are associated with divergent plate boundaries, such as midocean ridges and rift zones, include chemically precipitated copper, lead, and zinc sulfide deposits found within ocean floor sediments; hydrothermal vein and disseminated deposits of copper, lead, and zinc sulfides; and magmatic segregation deposits of chromium and copper and nickel sulfides

b. On the crest of the midocean ridge (a divergent plate boundary) in the Pacific ocean (the East Pacific Rise) are submarine hot springs known as *black smokers;* these springs spew hot black plumes of metallic sulfide minerals into cold seawater, which subsequently precipitates and forms mounds of metal deposits on the sea floor around the vent that resemble a chimney

c. Deposits that are associated with convergent plate boundaries where plates are subducted include contact metamorphic deposits, vein deposits, disseminated and porphyry copper deposits, and pegmatitic deposits

d. Deposits that are found in continental interiors (cratons) mark sites where continents collided (suture zones) and include chemically precipitated deposits rich in iron and magnesium, disseminated hydrothermal deposits, and magmatic segregation deposits; these deposits, which originally formed on the sea floor, were scraped off on the continent when plates converged and an ocean basin closed

D. Important metallic minerals

1. Dozens of metals are known and used, but the most important ones are iron, copper, aluminum, lead, zinc, gold, and silver
2. Iron is considered the most important metal of modern civilization
   a. About 95% of all metal consumed in the world is iron
   b. Iron is frequently alloyed (combined) with other metals to add properties of hardness, toughness, durability, and resistance to corrosion; several of the metals that alloy with iron include manganese, chromium, nickel, molybdenum, cobalt, vanadium, and tungsten
   c. The chief ores of iron are hematite (Fe_2O_3) and magnetite (Fe_3O_4)
   d. Iron is formed in magmatic segregation deposits of igneous origin, in banded iron deposits of sedimentary origin, and in laterites (a residue of the chemical weathering process)
3. Copper is another important metal
   a. Approximately 70% of the copper produced is used in the electrical industry
   b. Most copper ores are sulfides; the principal ones include native copper (Cu), chalcopyrite ($CuFeS_2$), and chalcocite (Cu_2S)
   c. Copper may form as an early crystallization mineral of a mafic magma or may be hydrothermal in origin
4. Aluminum is another widely used metal; aluminum is used in the manufacture of beverage cans, airplane structures, and many other products
   a. The chief ore of aluminum is bauxite (Al_2O_3 znH_2O), a mixture of hydrated aluminum oxide, which generally is obtained through open-pit mining
   b. Aluminum has a sedimentary origin; it is the end product of intense chemical weathering of aluminum-bearing minerals (primarily feldspars)
5. Lead primarily is used in the manufacture of batteries
   a. The chief ore of lead is galena (PbS)
   b. Lead is of hydrothermal origin, commonly associated with copper and zinc deposits
   c. Lead is obtained by both underground and open-pit mining

6. Zinc is another metal widely used in industry; it is used principally for galvanization and the manufacture of brass and other alloys
    a. The chief ore of zinc is the mineral sphalerite (ZnS)
    b. Zinc is found in association with lead; most lead mines also extract zinc
7. Gold belongs to the precious metal group
    a. Gold is used primarily in coins, jewelry, and dentistry
    b. Gold most often is found in the native state (uncombined with other elements) in the form of nuggets and grains; it also is found in hydrothermal veins and in placer deposits
8. Silver is another precious metal
    a. Silver is used in coins, jewelry, tableware, and many other products
    b. Silver is found in the native state and in sulfide ores; most silver is obtained as a byproduct of lead and copper mining

IV. Nonmetallic Resources

A. General information
1. Nonmetallic resources generally are not called ores, but are classified as industrial rocks and minerals; resources of this type include materials, such as stone used for building, road aggregate, clays used for ceramics and pottery, and fertilizers
2. Many nonmetallic resources occur naturally and can be used directly with little preparation (for example, sand and gravel), whereas others must undergo some type of preparation (for example, limestone in making cement)
3. Types of naturally occurring industrial minerals include rock salt, quartz, sulfur, and sylvite
4. Other types of nonmetallic resources include wood, groundwater, and soil

B. Types of nonmetallic resources
1. Some of the more widely used types of nonmetallic resources include sand and gravel, building stone, crushed stone, rock salt, gypsum, clay, and fertilizers
2. Sand and gravel are used in the preparation of concrete for highway and building construction; pure quartz sand is used to make window glass
3. Building stone, typically limestone or granite, is removed from quarries in blocks to construct buildings; crushed stone (mostly limestone) is used in forming roadbeds during highway construction
4. Rock salt, a coarse, crystalline halite, forms as an evaporite and can be obtained from underground mines or extracted from seawater; one of the primary uses of rock salt is table salt
5. Gypsum, a sedimentary evaporite mineral, is used to make plaster and wallboard
6. Clay, which is produced by the chemical weathering of common rock-forming minerals, is used in making ceramics
7. Fertilizers, which include phosphate, nitrate, and potassium compounds, are used in agriculture; phosphate is produced from deposits composed of the remains of certain marine organisms; nitrate and potassium can form directly as evaporite deposits

Study Activities

1. Describe the difference between a resource and a reserve.
2. Outline the three major classifications of natural resources, and explain at least one method by which each is extracted.
3. Compare the structural and stratigraphic traps used in drilling for petroleum, and describe the most common one.
4. List four metallic and four nonmetallic resources and explain how each is used in industry.
5. Select three criteria that determine whether a mineral resource can be considered an ore.
6. Define what is meant by renewable and nonrenewable resources.

Appendix

Selected References

Index

Appendix: Glossary

Absolute time—commonly refers to ages, in years, determined by radiometric dating; also refers to ages obtained from tree rings, varves, and the like

Age—division of earth's history of unspecified duration, marked by a dominance of a particular life-form (such as the Age of Fishes) or by special physical conditions (such as the Ice Age)

Alluvial fan—deposits formed in arid or semiarid regions where streams, which have been confined to canyons and arroyos, spread out in a fan shape onto a plain or valley floor

Angular unconformity—erosional surface that separates layers of tilted sedimentary rocks below from flat-lying sedimentary rock layers above it

Anticline—upwarp or arched portion of a fold

Asthenosphere—zone located in the upper mantle, below the lithosphere, at depths of about 80 to 125 km; area in the earth where rocks behave plastically and on which the crust of the earth is thought to float

Atom—smallest particle that exists as an element

Base level—level below which a stream can no longer erode its bed

Batholith—generally discordant igneous intrusive body which has more than 100 square kilometers of surface exposed

Bedrock—solid rock underlying soil, sand, and clay

Bowen's reaction series—process that explains the formation of intermediate and felsic magmas from a mafic (basaltic) magma

Cementation—precipitation of agents, such as silica and calcium carbonate, in pore spaces, resulting in binding together of sediment; this becomes rock

Cenozoic Era—subdivision of geologic time (derived from the Greek word for recent life), in which the mammals rose to prominence (Age of Mammals); follows the Mesozoic Era, and began about 65 million years ago and extends to the present

Chemical weathering—process of weathering by which water, carbon dioxide, and oxygen transform rocks and minerals into new chemical combinations that are stable under prevailing conditions at or near the earth's surface

Clastic—sedimentary rock texture composed of fragments of preexisting rocks and minerals formed by mechanical weathering

Coal—readily combustible sedimentary rock resulting from compaction and induration of altered plant remains (similar to those in peat); classified by the different kinds of plant material (type), the degree of metamorphism (rank), and range of impurities (grade)

Compaction—pressure from the weight of overlying sediment reduces the volume of underlying sediment layers

Concordant—contact between igneous rock bodies that parallel the bedding or foliation of the country rock into which it intrudes

Contact metamorphism—metamorphism that is produced by igneous intrusion; although it is primarily thermal, some deformation of country rock may result

Continental drift—theory that the present continents have been formed by the earlier breakup of one large continent and have drifted to their current positions

Convection cells—mass movement of material where hotter material rises and then sinks because of differences in temperature and density; they are thought to exist in the mantle

Convergent plate boundary—area where two plates are colliding (also called convergent plate margins)

Core—central part of the earth thought to consist primarily of iron and nickel

Correlation—matching of rocks of a particular age that are found in one place with other rocks found elsewhere

Crust—outermost, rigid portion of the earth's surface consisting of continental and oceanic crustal material; it is less dense than the mantle beneath it

Deflation—removal of material from a desert, beach, or other land area by wind

Diastrophism—movement of the earth's crust caused by tectonic processes; produces mountains and basins

Differentiation—process by which different types of igenous rocks form from cooling magma

Dike—tabular body of igneous rock that is discordant to country rock into which it intrudes

Dip—angle or tilt of strata or other surface from the horizontal; it is measured perpendicular to the strike

Disconformity—uncomformity in which the beds above and below the plane of unconformity are parallel to each other

Divergent plate boundary—region where two plates are moving away from each other (such as the mid-Atlantic ridge); also called divergent plate margins

Drainage basin—area drained by a stream and its tributaries

Earthquake—sudden or "shaking" movement of part of the earth's crust caused by the sudden release of energy in the earth's crust or mantle

Elastic rebound theory—theory that explains how rocks "snap back" into their original shape after stored energy is released

Emergent coastline—coastline that results from a fall in sea level or rise in land; commonly has marine terraces with beaches

Eon—largest division of geologic time; subdivided into eras

Epeirogenic—refers to the upward or downward movement of large areas of the earth's crust without folding

Epicenter—point on the earth's surface directly above the focus of an earthquake

Epoch—division of geological time smaller than a period

Era—division of geologic time; it is a subdivision of an eon

Erosion—weathering away of rocks by mass wasting, wind, water, and glaciers

Exfoliation—process by which the outer surface of a rock is broken off in concentric slabs; occurs in granite and other igneous rocks

Extrusive—refers to rocks that have formed from lava flows or pyroclastic material erupted from the earth's surface; commonly known as volcanic rocks

Fault—a break in the rocks along which movement has taken place; it typically is parallel to the fracture

Fault scarp—scarp formed by an uplifted fault that reaches the earth's surface

Focus—place in the earth where an earthquake actually occurs

Folded mountain—mountain produced by large-scale folding or shortening of the crust

Footwall—mass of rock beneath an inclined fault

Fossil—organic remains, trace, or imprint of an animal or plant preserved in rock

Fossil succession—progression of life from a simple to a more complex form through time; these remains create a fossil record

Geologic time scale—chronological organization that correlates rock strata to the history of the earth; recorded in the succession of rocks

Geothermal energy—energy derived from the earth's internal heat

Geyser—spring that intermittently ejects hot water and steam (when hot groundwater cannot circulate freely)

Glacier—large mass of ice formed on land by the compaction of successive layers of snow; moves slowly downhill due to the stress of its own weight

Graben—narrow block of the earth's crust that has moved downward between normal faults on either side of it

Graded bedding—bedding in which the largest particles are at the bottom of a layer and the smallest particles are at the top

Groundwater—water that fills the pore spaces in rocks and soil

Half-life—Time period in which half the initial number of atoms of a radioactive element break up into atoms of another element or isotope

Hanging valley—Smaller valley that terminates above a main valley; typically results from glacial erosion

Hanging wall—rock that lies above an inclined fault

Horst—uplifted block bounded by normal faults on either side of it

Hydrologic cycle—constant circulation of water from the sea, through the atmosphere, to the land, and its return to the sea

Igneous rocks—crystalline, or glassy, rocks that have solidified from magma

Index fossils—fossils that clearly establish the age of strata in which they are found

Intensity—refers to the effects of an earthquake, that is, the damage done to people and buildings

Intrusive —refers to rock formed by emplacement of magma into preexisting rock

Island arcs—curved chains of andesitic volcanoes rising from the sea floor; typically located between an oceanic trench and a continent

Isostasy—theory stating that different masses of the earth's crust stand in gravitational balance or equilibrium with each other and within the earth

Isotope—element that has more than one atomic structure; all structures have the same atomic number but vary in atomic mass (that is, the number of neutrons)

Laccolith—an igneous intrusion with a flat floor and a rounded roof which has pushed sediments above it into the shape of a dome

Lithosphere—comprises the outer, solid part of the earth consisting of the crust and the upper part of the mantle; has a depth of about 100 km

Lithostatic pressure—vertical pressure caused by the weight of overlying rock

Magma—molten rock material generated in the earth's upper mantle at depths of 50 to 200 km

Magnitude—measure of energy released during an earthquake; measured by the Richter scale

Mantle—middle layer of rock material that separates the earth's crust above from the core below

Mantle plume—Narrow column of hot mantle rock that rises and spreads out radially; it breaks up the lithosphere and causes the plates to move, thereby producing a hot spot at the earth's surface

Mass movement—movement of bedrock, rock debris, or soil downward that is caused by gravity; also called mass wasting

Mechanical weathering—breakdown of rock into smaller pieces by weathering or other physical process

Mercalli scale—scale that measures the intensity of an earthquake

Mesozoic Era—follows the Paleozoic Era and precedes the Cenozoic Era on the geologic time scale; also known as the Age of Reptiles, of which dinosaurs are the best known

Metallic minerals—class of elements that generally is characterized by metallic luster, malleability, and conductivity of heat and electricity

Metamorphic grade—the intensity of metamorphism that indicates the pressure and temperature conditions (or facies) in which the metamorphism took place

Metamorphic rocks—rocks that have undergone mineralogical, chemical, or structural changes

Metamorphism—transformation of preexisting solid rocks by heat, pressure, or chemically active solutions

Mohorovicic discontinuity—boundary that separates the crust from the mantle

Noncomformity—an erosional surface between an igneous or metamorphic rock and overlying sedimentary rock strata

Normal fault—fault in which the hanging wall moves downward relative to the footwall

Ore—rock material containing naturally occurring mineral that can be profitably mined

Original horizontality—the concept that sedimentary rocks were originally deposited in flat, horizontal layers (strata)

Orogenic—refers to mountain building typically accompanied by intrusion of igneous rocks and by folding

Paleomagnetism—study of remnant magnetism in ancient rocks to determine the intensity and direction of the earth's magnetic field in the past

Paleozoic Era—portion of geologic time in which complex life first appeared; precedes the Mesozoic Era

Period—subdivision of an era; part of geologic time during which a sequence of rocks designated as a system was deposited

Permeability—capacity of rock or other material to transmit liquid or gas

Phanerozoic Eon—refers to the time of visible life; follows the Proterozoic Eon on the geologic time scale and comprises the Paleozoic, Mesozoic, and Cenozoic Eras

Physical geology—study of the earth, changes that occur at the surface and in the interior of the earth, and the forces that cause those changes

Plate tectonics—theory proposing that earth's surface is covered by a number of thin plates (the lithosphere) which move over the underlying material (the asthenosphere)

Plutonic rocks —igneous rocks formed by magma which solidified before reaching the earth's surface; all plutonic rocks are intrusive rocks

Polar wandering—theory that the earth's magnetic poles have changed position throughout geologic time

Porosity—the amount of water a rock can hold, which depends on the volume of its pore spaces

P wave—high-frequency compressional wave which travels in the same direction that the movement (such as the vibration of rock) travels

Pyroclastic—ejected, broken fragments of material resulting from the sudden eruption of viscous magma

Radiocarbon dating—dating method that uses the rate of decay of carbon 14 to carbon 12; used primarily to date organic remains up to 70,000 years before the present

Recrystallization—when original mineral grains slowly dissolve and typically form larger crystals

Regolith—loose rock material produced by weathering that rests on bedrock

Relative time—sequence of events arranged relative to one another; not measured in years

Reserve—all resources from which a mineral or fuel can be extracted profitably with existing technology

Resource—all known, worldwide deposits of geologic materials, including those not yet discovered

Richter scale—measure of the intensity of an earthquake; ranges from 0 to 8.5

Runoff—water falling on the earth's surface that reaches streams or rivers

Saltation—process whereby particles of sand or sediment are transported in a series of rolls or bounces

Sea-floor spreading—hypothesis stating that new oceanic crust is being formed by rising basaltic magma at the midocean ridges and is moving away from the ridges at a rate of 1 to 10 cm per year

Sedimentary rocks—rocks formed from lithification or weathering of preexisting rocks

Seismograph—an instrument designed to detect, measure, and record the earth's vibrations

Sill—an intrusion concordant with the country rock; it generally is horizontal

Sinkhole—surface depression caused by the collapse of a cavern roof or by the dissolution of limestone

Soil—unconsolidated, weathered rock combined with water, air, and organic matter

Solifluction—flow of wet material (at the surface) which takes place when the ground in a permafrost area is partly thawed

Stalactites—calcium carbonate (calcite) deposit hanging from a cave ceiling

Stalagmites—calcium carbonate (calcite) deposit standing up from a cave floor

Stratification—the arrangement of sedimentary rocks in horizontal layers, called beds

Strike—formed by intersection of an inclined plane with a horizontal plane; measured by compass direction

Submergent coastline—coastline created because the sea level has risen or the land has subsided

Superposition—the order in which rocks are placed one above the other, with the youngest on top

S wave—high-frequency seismic wave which travels perpendicular to the direction that movement travels

Syncline—a downwarp or trough portion of a fold that typically is U-shaped

Texture—relationship between mineral grains in a rock; refers to the size, shape, and arrangement of particles or crystals in rock

Thrust fault—a fault with a 45° or less dip, in which the hanging wall appears to have moved upward relative to the footwall; also called a reverse fault

Tsunami—a large wave caused by an oceanic earthquake or volcanic eruption which causes displacement of the sea floor; also called seismic sea wave

Turbidity current—a mass of water carrying sediment that travels swiftly and violently down a subaqueous slope

Unconformity—rock strata that contains no sediments from a particular geologic period (perhaps resulting from nondeposition or erosion); occurs between sedimentary rocks and rocks on which they rest

U-shaped valley—formed by glaciers moving through, straightening, and widening an old river valley

Water table—upper limit of the zone of saturation, where the pore spaces are saturated with water

Weathering—process by which rocks at or near the earth's surface undergo physical breakdown and chemical decomposition

Selected References

Bates, R.L., and Jackson, J.A. (eds.). *Glossary of Geology.* Alexandria, Va.: American Geographical Institute, 1990.

Blackburn, W.H., and Dennen, W.H. *Principles of Mineralogy.* Dubuque, Iowa: William C. Brown, 1988.

Foster, R.J. *Geology* (5th ed.). Columbus, Ohio: Charles E. Merrill Publishing Co., 1985.

Monroe, J.S., and Wicander, R. *Physical Geology: Exploring the Earth.* St. Paul: West Publishing Company, 1992.

Plummer, C.C., and McGeary, D. *Physical Geology* (6th ed.). Dubuque, Iowa: William C. Brown, 1993.

Skinner, B.J., and Porter, S.C. *The Dynamic Earth: An Introduction to Physical Geology* (2nd ed.). New York: John Wiley & Sons, 1992.

Tarbuck, E.J., and Lutgens, F.K. *Earth Science* (6th ed.). New York: Macmillan, 1991.

Thompson, G.R., and Turk, J. *Earth Science and the Environment.* Philadelphia: W.B. Saunders, 1993.

Tucker, M.E. *Sedimentary Petrology: An Introduction to the Origin of Sedimentary Rocks* (2nd ed.). Oxford, England: Blackwell Scientific Publications, 1991.

Wincander, R., and Monroe, J.S. *Historical Geology: Evolution of the Earth and Life through Time* (2nd ed.). St. Paul: West Publishing Company, 1993.

148

Index

A
Absolute time, 9, 55, 57
Acid rain, 34
Agassiz, Louis, 90
Ages, geologic, 10, 58
Alluvial fan, 42, 70, 97
Aluminum, 14t, 138
Amphibole, 19i, 20, 24i
Andesite, 27
Anion, 13
Anomalies, 117, 124
Anthracite, 46
Anticline, 119, 120i, 121
Aquifer, 80, 82
Archean Eon, 10, 58
Arête, 87, 88i
Artesian well, 80
Asthenosphere, 2, 3i, 115, 129
Atoll, 107i
Atom, 11-13
Augite, 20
Avalanche, 63

B
Bajada, 97
Barrier island, 102i
Basalt, 27
Basin, 73, 119
Batholith, 25-26
Baymouth bar, 101, 102i
Beaches, 102i, 103
Bedrock, 35, 36i, 62
Benioff zone, 110, 131
Bergschrund, 86
Berm, 103
Biotite, 20, 24i
Bonding, chemical, 12
Bowen's reaction series, 24i, 25
Breccia, 44, 122
Brongniart, Alexandre, 9
Butte, 98

C
Calcium, 14t
Calderas, 29
Caliche, 37
Cambrian Period, 9, 58-59
Carbonation, 34
Carboniferous Period, 60
Cation, 13
Caves and caverns, 82-83
Cementation, 43
Cenozoic Era, 10, 60
Chalk, 45
Channel, 67
Chemicals
  bonding of, 12
  weathering and, 32, 34

Chert, 45
Cirque, 86, 87, 88i
Clastics, 39-40, 43
Clay, 19i, 20, 139
Claystone, 45
Cleavage, 16
Climate
  glaciers and, 91
  groundwater and, 78
  landform development and, 75
  mass movement and, 65
  soil production and, 35
  weathering and, 32
Coal, 45-46, 134-136
Coastlines, 103-105
Compaction, 43
Condensation, 8i
Conglomerate, 44
Continental crust, 2, 3i, 114, 125
Continental divide, 73
Continental drift, 5, 127-128
Continental shelf, 106
Continuity, lateral, 2
Convection cell, 4, 132
Copper, 138
Core, 2, 3i, 115
Correlation, 55
Creep, 63
Cretaceous Period, 60
Crevasse, 86
Cross-bedding, 46, 96
Cross-cutting, 2
Crust
  continental, 2, 3i, 114, 125
  elements in, 14t
  oceanic, 2, 3i, 105, 128-129
Crystal, structure of, 16, 17i
Crystallization, 6i, 24i
Currents, 46, 100-101, 106
Cuvier, Georges, 9

D
Dating
  absolute, 55, 57
  radiocarbon, 57
  relative, 54-55
Davis, W.M., 75
Decarbonation reaction, 49
Deflation, 93, 94
Deformation, 118-122
Dehydration reaction, 49
Delta, 70
Deposition
  by glaciers, 89-90
  rock cycle and, 6
  sediments and, 42-43
  by streams, 70

Deposition *(continued)*
  by waves, 101
Desert, 96-98
Devonian Period, 59-60
Diastrophism, 49, 118
Diatomite, 45
Differentiation, 23, 25
Diffusion, 49
Dikes, 26
Diorites, 27
Dip, 121
Disconformity, 55, 56i
Discontinuity, 114
Dolostone, 45
Dome, 119, 120i
Drainage, 73, 74i, 75
Drumlin, 89i, 90
Dunes, 94, 95i

E
Earth
  composition of, 114-115
  continental crust of, 2, 3i, 114, 125
  elements of, 13, 14t
  magnetic field of, 116-117
  oceanic crust of, 2, 3i, 105, 128-129
  resources of, 133-139
  rock types of, 3-4
  structure of, 2, 3i
  temperature of, 115-116
Earthflow, 63
Earthquakes
  causes of, 110, 111i
  distribution of, 109-110
  effects of, 112-113
  seismic waves and, 110-112
Elastic rebound theory, 109, 110, 111i
Electron, 11, 12
Elements, 13, 14t
End members, 13-14
Energy
  earthquakes and, 111i
  groundwater and, 83
  resources for, 134-136
Eocene Epoch, 60
Eons, 10, 58
Epicenter, 112
Epochs, 10, 58
Eras, 10, 58
Erosion
  by glaciers, 87
  landform development and, 75
  rock cycle and, 6
  by streams, 69, 72
  by waves, 101

i refers to an illustration; t, to a table

Erosion *(continued)*
  by weathering, 31
  by wind, 94
Esker, 89i
Estuary, 103
Evaporation, 8i
Exfoliation, 32

F
Facies, metamorphic, 52
Fall, 64i, 65
Fault, 106, 109, 121-122, 124i
Fault scarp, 122
Feldspars, 17i, 19i
  plagioclase, 18, 24i
  potassium, 18, 20, 24i
  weathering and, 34
Fertilizer, 139
Firn, 85
Fjord, 87
Flood plain, 70
Flow, 63, 64i
Focus, 112
Folds, 119, 120i, 121
Fossil, 2, 55, 128
Fracture, 16
Frost, 32

G
Gabbros, 27
Galena, 17i
Gas, natural, 134-136
Geodes, 83
Geology
  absolute dating and, 55, 57
  principles of, 1-2
  relative dating and, 54-55
  time scale and, 7, 9-10, 57-
    58, 59t, 60
Geothermal gradient, 22, 49,
    115-116
Geyser, 83
Glaciers
  deposition and, 89-90
  erosion and, 87
  formation of, 85
  glaciation and, 90-92
  landforms and, 87, 88i, 89i
  movement of, 86
  sediments and, 41-42
  types of, 86-87
Gneiss, 51
Gold, 139
Graben, 106, 123, 124i
Grade, metamorphic, 52
Graded bedding, 46
Granites, 27
Graywacke, 45
Greenhouse effect, 91
Groundwater, 8i, 67
  availability of, 78

Groundwater *(continued)*
  caves and, 82-83
  geothermal energy and, 83
  movement of, 80
  outlets for, 80
  pollution and, 80, 82
  spring formation and, 81i
  subsurface zones of, 78,
    79i, 80
Gutenberg, Beno, 113-114
Guyot, 107
Gypsum, 17i, 45, 149

H
Half-life, 57
Hardness scale, minerals
    and, 15t
Hess, Harry, 5, 127-128
Holocene Epoch, 60
Horizontality, original, 2, 55
Horn, 87, 88i
Hornblende, 20
Hornfels, 51
Horst, 123, 124i
Hot springs, 83
Humus, 35
Hutton, James, 1-2, 54
Hydration, 34
Hydroelectricity, 134
Hydrologic cycle, 4, 7, 8i
Hydrolysis, 34

I
Ice, 85-92
Ice Age, 60, 90-91
Igneous rocks
  classification of, 26t, 27
  composition of, 27
  description of, 3
  extrusive, 28-29
  intrusive, 25-26
  plutonic, 25
  texture of, 26-27
Index fossils, 55
Index minerals, 48
Inselberg, 97
Ion, 12-13
Iron, 14t, 138
Island arc, 107
Isostasy, 103, 123-125
Isotope, 11, 55, 57t

J
Jetty, 105
Joints, 121
Jurassic Period, 60

K
Kame, 89i, 90
Karst topography, 82
Kettle, 89i, 90

L
Laccolith, 26, 123
Lake
  oxbow, 70, 71i
  playa, 97
  pluvial, 91
  proglacial, 90
Landscape
  glaciers and, 88i, 89i
  streams and, 72-75
Landslide, 63
Laterites, 37
Lattice, crystal, 13, 17i
Lava, 21, 28-29
Lavoisier, Antoine, 9
Leaching, 35, 36i, 37
Lead, 138
Levee, 70
Lignite, 46
Limestone, 45
Liquefaction, 63, 113
Lithification, 6i
Lithosphere, 2, 3i, 115, 129
Loam, 37
Loess, 93, 94
Luster, minerals and, 16

M
Magma
  Bowen's reaction series
    and, 24i
  composition of, 22-24
  consolidation of, 24-25
  description of, 21-22
  formation of, 22
  rock cycle and, 6i
  volcanoes and, 28-29
Magnesium, 14t
Magnetic field, 116, 129
Mantle, 2, 3i, 114
Mantle plumes, 21, 132
Marble, 51
Mass movement, 62-63, 64i,
    65-66
Meander, 70, 71i
Mercalli scale, 112
Mesa, 98
Mesozoic Era, 10, 60
Metals, bonding of, 12
Metamorphic rocks
  classification of, 50, 51t
  composition of, 51
  description of, 4
  grades of, 48, 52
  texture of, 50-51
Metamorphism, 6i, 48-50
Meteorites, 3, 115
Mica, 19i, 20
Microcline, 20
Migmatite, 48
Milankovitch, Milutin, 91

i refers to an illustration; t, to a table

Stream *(continued)*
   discharge of, 68
   drainage and, 73, 74i, 75
   erosion by, 69
   nature of, 67
   oxbow lakes and, 71i
   transport and, 69
   valleys and, 72-73
   velocity of, 68
Stress, 118-119
Strike, 121
Subsoil, 37
Succession, fossil, 2, 55
Superposition, 2, 55
Syncline, 119, 120i, 121

T
Talus, 98
Tarn, 87, 88i
Temperature
   of Earth's interior, 115-116
   metamorphism and, 49
Tertiary Period, 60
Tetrahedron, silica, 16, 18i, 19i
Texture
   of igneous rocks, 26-27
   of metamorphic rocks, 50-
      51
   of sedimentary rocks, 43
Tides, 104-105
Till, 89i
Time, geologic, 7, 9-10, 54-60
Tombolo, 101, 102i
Topsoil, 37
Transpiration, 8i
Transportation
   rock cycle and, 6
   sediments and, 41-42
Travertine, 83
Triassic Period, 60
Tsunami, 99, 113
Turbidity current, 46, 106

U
Unconformity, 55, 56i
Uniformitarianism, 1-2, 54
Uranium, 134, 136
Ussher, Archbishop James, 2

V
Valence, 13
Valleys
   glaciers and, 87, 88i
   streams and, 72-73
van der Waal's bonding, 12
Varve, 90
Ventifact, 94
Viscosity, magma and, 22
Volcanoes, 28-29, 107-108,
   123

W-X
Water. *See also* Groundwater;
   Ocean
   geothermal, 83
   mass movement and, 65
   sediments and, 41-42
   streams and, 67-75
Water table, 78, 79i, 80
Waves
   ocean, 99-100, 101i, 102
   seismic, 110-112
Weathering
   chemical, 32, 34
   description of, 31
   effects of, 34
   mass movement and, 65
   mechanical, 32, 33i
   rock cycle and, 5, 6i
   soil production and, 35
Wegener, Alfred, 5, 127-128
Wells, 80
Wind
   definition of, 93
   deposition and, 94-96
   erosion by, 94
   sand dunes and, 95i
   sediments and, 41-42, 94
Wood, petrified, 83

Y-Z
Yardang, 94
Zinc, 139
Zircon, 17i
Zone of leaching, 35, 36i

i refers to an illustration; t, to a table

Notes